Contributions to Management Science

Frank Meisel

Seaside Operations Planning in Container Terminals

Physica-Verlag

A Springer Company

Frank Meisel
Martin-Luther-Universität Halle-Wittenberg
Juristische und
Wirtschaftswissenschaftliche Fak.
Große Steinstr. 73
06108 Halle
Germany
frank.meisel@wiwi.uni-halle.de

ISSN 1431-1941
ISBN 978-3-7908-2190-1 e-ISBN 978-3-7908-2191-8
DOI 10.1007/978-3-7908-2191-8
Springer Dordrecht Heidelberg London New York

Library of Congress Control Number: 2009926247

Cover design: WMXDesign GmbH, Heidelberg

Printed on acid-free paper

Springer is part of Springer Science+Business Media (www.springer.com)

Acknowledgments

This book is the result of my research carried out at the Faculty of Law and Economics at Martin-Luther-University Halle-Wittenberg, Germany, from 2004 to 2008. This work could not have been accomplished without the support of several persons.

First of all, I would like to express my deepest gratitude to Prof. Dr. Christian Bierwirth, holder of the Chair of Production and Logistics at Martin-Luther-University, for pointing me to the interesting subject of container terminal operations planning, for supervising my research and for providing a constant source of inspiration and motivation throughout the years.

I would like to thank Prof. Dr. Hans-Otto Günther for continuous discussion and for co-refereeing this thesis. Prof. Dr. Kap Hwan Kim and Dr. Young-Man Park have supported my research by kindly providing their benchmark suites.

Many thanks go to all my colleagues who provided helpful comments on my research at conferences, in projects and in personal discussions. I would like to thank particularly the team of the Chair of Production and Logistics for their support, namely Susanne Berger, Angela Herrmann and Ute Lorenz.

My final thanks go to my whole family, all my friends and in particular to Adrienne. You have unconditionally helped and encouraged me all the time throughout this endeavor. Thank you very much.

Halle
November 2008

Frank Meisel

Contents

List of Figures

List of Tables

Notations

Notation of the berth allocation and quay crane assignment study in Chap. 6:

n	Number of vessels to be served
V	Set of vessels to be served, $V = \{1,2,\ldots,n\}$
Q	Number of available QCs at the terminal
L	Number of 10-m quay segments (length of the quay)
T	Set of 1-h periods, $T = \{0,1,\ldots,H-1\}$, H is the planning horizon
l_i	Length of vessel $i \in V$ given as a number of 10-m segments
b_i^0	Desired berthing position of vessel i
m_i	Crane capacity demand of vessel i given as a number of QC-hours
$r_i^{\min}$	Minimum number of QCs agreed to serve vessel i simultaneously
$r_i^{\max}$	Maximum number of QCs allowed to serve vessel i simultaneously
R_i	Feasible range of QCs assignable to vessel i, $R_i = [r_i^{\min}, r_i^{\max}]$
ETA_i	Expected time of arrival of vessel i
EST_i	Earliest starting time if vessel i speeds up on journey, $EST_i \le ETA_i$
EFT_i	Expected finishing time of vessel i
LFT_i	Latest finishing time of vessel i without penalty cost arising
c_i^1, c_i^2, c_i^3	Service quality cost rates for vessel i given in units of 1,000 USD per hour
c^4	Operational cost rate given in units of 1,000 USD per QC-hour
α	Interference exponent
β	Berth deviation factor
$d_i^{\min}$	Minimum handling time needed to serve vessel i
r_i^{lvl}	Resource level of vessel i
P	Priority list of vessels
M	A large positive number
b_i	Integer decision variable, berthing position of vessel i
s_i	Integer decision variable, time of starting the handling of vessel i, i.e., the berthing time of the vessel

e_i	Integer decision variable, time of ending the handling of vessel i, i.e., the finishing and departure time of the vessel
r_{it}	Binary decision variable, set to 1 if at least one QC is assigned to vessel i at time t
r_{itq}	Binary decision variable, set to 1 if exactly q QCs are assigned to vessel i at time t, $q \in R_i$
Δb_i	Integer decision variable, deviation between the desired and the actually chosen berthing position of vessel i, $\Delta b_i = \lvert b_i^0 - b_i \rvert$
ΔETA_i	Integer decision variable, required speedup of vessel i to reach its berthing time, $\Delta ETA_i = (ETA_i - s_i)^+$
ΔEFT_i	Integer decision variable, tardiness of vessel i, $\Delta EFT_i = (e_i - EFT_i)^+$
u_i	Binary decision variable, set to 1 if the finishing time of vessel i exceeds LFT_i
y_{ij}	Binary decision variable, set to 1 if vessel i is berthed below of vessel j
z_{ij}	Binary decision variable, set to 1 if service of vessel i ends not later than service of vessel j starts

Notation of the quay crane scheduling study in Chap. 7:

n	Number of loading and unloading tasks for a vessel
Ω	Set of loading and unloading tasks, $\Omega = \{1,2,\ldots,n\}$
$0, T$	Dummy tasks, $T = n+1$
$\Omega^0, \Omega^T, \overline{\Omega}$	Task sets, $\Omega^0 = \Omega \cup \{0\}, \Omega^T = \Omega \cup \{T\}, \overline{\Omega} = \Omega \cup \{0,T\}$
p_i	Processing time of task $i \in \overline{\Omega}$, $p_0 = p_T = 0$
l_i	Bay position of task $i \in \Omega$
Φ	Set of precedence constrained task pairs
Ψ	Set of task pairs forbidden to be processed simultaneously, $\Psi \supseteq \Phi$
q	Number of QCs assigned to a vessel
Q	Set of QCs numbered by increasing initial bay positions, $Q = \{1,2,\ldots,q\}$
r^k	Ready time of QC $k \in Q$
l_0^k	Initial bay position of QC k
$\hat{t}$	Crane travel time to traverse between adjacent bays, $\hat{t} > 0$
t_{ij}	Crane travel time to traverse between bays l_i and l_j, $t_{ij} = \hat{t}\lvert l_i - l_j \rvert$
t_{0j}^k	Travel time of QC k to traverse from its initial bay position to bay l_j, $t_{0j}^k = \hat{t}\lvert l_0^k - l_j \rvert$
δ	Safety margin, the minimum number of in-between bays that has to be kept between two adjacent QCs at all times
δ_{vw}	The smallest allowed difference between the bay positions of cranes v and w, $\delta_{vw} = (\delta + 1)\lvert v - w \rvert$
Δ_{ij}^{vw}	Minimum temporal distance between the completion of task i processed by QC v and task j processed by QC w
Θ	Set of combinations of tasks and QCs for which a positive minimum temporal distance must be ensured

m_i	Crane to which task i is assigned
M	A large positive number
x_{ij}^k	Binary decision variable, set to 1 if tasks i and j are consecutively processed by QC k
z_{ij}	Binary decision variable, set to 1 if task j starts after task i is completed
c_i	Integer decision variable, completion time of task $i \in \overline{\Omega}$

Notation of the quay crane scheduling with crane time windows study in Chap. 7:

τ_k	Number of time windows of QC k
TW_k	Set of time windows of QC k, $TW_k = \{1,\ldots,\tau_k\}$
r^{ku}	Crane ready time of time window $u \in TW_k$
d^{ku}	Crane withdraw time of time window u
l_0^{ku}	Initial crane position of time window u
l_T^{ku}	Final crane position of time window u
t_{0i}^{ku}	Travel time of QC k to traverse from the initial crane position of time window u to the bay position of task i, $t_{0i}^{ku} = \hat{t}\vert l_0^{ku} - l_i\vert$
t_{iT}^{ku}	Travel time of QC k to traverse from the bay position of task i to the final crane position of time window u, $t_{iT}^{ku} = \hat{t}\vert l_T^{ku} - l_i\vert$
y_i^{ku}	Binary decision variable, set to 1 if task i is processed by QC k in $u \in TW_k$

Notation of the integration study in Chap. 8:

U_{iq}	Average utilization rate of q cranes serving vessel i
$M_1,\ldots,M_6$	Measures for assessing differences between two BACAP solutions

Abbreviations

AGV	Automated guided vehicle
ALV	Automated lifting vehicle
ARE	Average relative error
BACAP	Berth allocation and crane assignment problem
BAP	Berth allocation problem
CT	Container terminal
FCFS	First-come first-served
GA	Genetic algorithm
GRASP	Greedy randomized adaptive search procedure
LB	Lower bound
MAS	Multi-agent system
MIP	Mixed integer programming
OR	Operations research
QC	Quay crane
QCAP	Quay crane assignment problem
QCSP	Quay crane scheduling problem
QCSPTW	Quay crane scheduling problem with time windows
RCPSP	Resource constrained project scheduling problem
RE	Relative error
RMGC, RTGC	Rail mounted gantry crane, rubber tired gantry crane
SWO	Squeaky wheel optimization
TEU	Twenty-foot equivalent unit
TS	Tabu search
TW	Time window
UDS	Unidirectional scheduling heuristic
UDSTW	Unidirectional scheduling heuristic for the QCSPTW
USD	US dollar
VRP	Vehicle routing problem
YC	Yard crane

Chapter 1
Introduction

1.1 Motivation and Scope of Research

Container terminals in seaports constitute interfaces between sea and land transport of goods in global transport chains. These logistics facilities face an increasing demand of service capacity, as is reflected by a tremendous growth in the worldwide container transshipments per year. For example, the top 20 terminals in the world showed an average relative increase of 14% with respect to the number of handled container units from 2006 to 2007, see Port of Hamburg Marketing (2008).

In spite of this development, competition is high among container terminals within the same region. A terminal's customers, first and foremost the vessel operators, expect a high level of service quality where reliability is one of the most important dimensions, see Wiegmans et al. (2001). Regarding the service of a vessel, reliability means to realize all transshipment operations within its projected service time interval. The reliability of terminal operations impacts the reliability of vessels in meeting their liner schedules. According to Notteboom (2006) unexpected waiting times of vessels before berthing and unexpected low transshipment productivity at terminals are responsible for about 86% of liner schedule disturbances, see Fig. 1.1. Currently, many terminal operators counteract this situation by extending their transshipment capacities. They build new terminals or enlarge existing terminals and purchase new or upgrade existing equipment. Ilmer (2005) provides an overview of current projects for building terminal capacity in northern Europe. The outlined projects promise a doubling of transshipment capacity from the year 2004 to the year 2010, but environmental and socio-economic issues often limit these ambitious expansion plans. For terminals that are unable to realize capacity building investments, the only alternative to enhance service quality is to increase the productive utilization of the existing resources.

This thesis investigates three operations planning problems that deal with the utilization of the seaside terminal resources, namely the quay space and the quay cranes. The addressed problems are referred to as the *Berth Allocation Problem* (BAP), the *Quay Crane Assignment Problem* (QCAP), and the *Quay Crane*

F. Meisel, *Seaside Operations Planning in Container Terminals*,
Contributions to Management Science,
DOI: 10.1007/978-3-7908-2191-8_1, © Physica-Verlag Berlin Heidelberg 2009

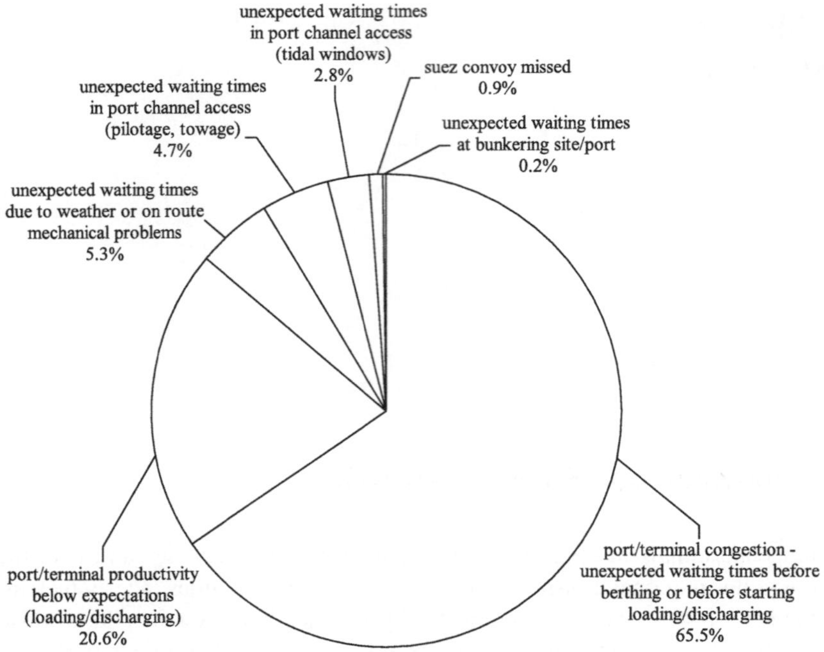

Fig. 1.1 Sources of liner schedule unreliability (survey data of East Asia – Europe relations, *source:* Notteboom, 2006)

Scheduling Problem (QCSP). Unfortunately, the scope, meaning, and distinction of these problems is not unique in the scientific literature. To cope with this difficulty, the problems are structurally distinguished in this thesis by the type of decisions made:

BAP: It determines a berthing time and a berthing position at the quay for each vessel to be served within a given planning horizon.
QCAP: It determines a set of specific cranes for the service of each vessel.
QCSP: It determines work schedules for the quay cranes serving a vessel.

The importance of these decision fields regarding the reliability of transshipment services is obvious. The berthing time decision determines the waiting time of a vessel. All decisions regarding the quay crane operations affect the transshipment productivity, which, in turn, determines the handling time and the departure time of a vessel. Furthermore, the utilization of cranes determines the dynamic labor force demand of the terminal and thus, the labor cost. Therefore, providing reliable services at suitable cost depends on the ability to solve the considered planning problems effectively. In practice the three problems are usually solved sequentially in the order of berth planning, crane assignment, and crane scheduling. As a consequence, the problems are often considered independently of each other in the scientific literature.

In this thesis an integrated solution approach to seaside planning problems is developed. The driving force of the approach is to anticipate vessel handling times on the basis of crane assignment and crane scheduling decisions. These specified handling times are used within the berth planning to decide on the berthing times and berthing positions of vessels. The following observation motivates such a problem integration. In the sequential solution process the impact of the crane resource on the vessel handling times is ignored. Hence, handling times must be estimated for the berth planning. Overestimating the handling time of a vessel, however, wastes quay space because the space is reserved for a longer time span than needed for the vessel's service. On the other hand, underestimating handling times leads to infeasible berth plans whenever a vessel is projected to berth although another vessel's service is not yet completed. These strong interrelations between the utilization of the quay space resource and the quay crane resource necessitate an integrated consideration of the problems.

The investigation aims at the design of a powerful integration concept. It has to be decided whether planning problems are integratively solved by merging them into a combined problem, or by enabling an interaction in terms of data exchange and a well defined solution order. Mathematical models have to be formulated to specify the optimization problems. In general, the resulting models use mixed integer formulations and thus, cannot be solved for instances of practical size. Hence, heuristic solution methods need to be developed. Since seaside operations planning has a strong impact on the competitiveness of a container terminal, objectives pursued in the planning models have to incorporate service quality and operational cost of a terminal. Moreover, the terminal management should be enabled to flexibly adapt the objectives to a current market situation. The performance of the solution methods and the contribution of integrated seaside operations planning to improved terminal service quality has to be assessed by computational tests.

1.2 Outline

Chapter 2 provides an introduction to the maritime container transport industry regarding its technical and organizational characteristics and trends. It gives furthermore insight into the layout of a container terminal and the different types of equipment used by the operations. Chapter 3 informally describes the operational planning problems of a container terminal. Thereby, the focus is put on seaside problems.

A large number of optimization approaches to the seaside problems has been published in the scientific literature. This motivates two classification schemes for the various problem variants. These schemes are provided in Chap. 4 and used for structuring the relevant literature in the field. Finally, the similarities between seaside operations planning problems and standard OR problems are discussed briefly.

The interrelations of the focused planning problems are explained in detail in Chap. 5. Next to the sequential solution process, which is nowadays applied in most terminals, theoretical concepts for the integration of planning problems are sketched. The range of integration concepts is critically assessed for planning seaside operations. This leads to the design of a new three-phase integration concept.

Chapter 6 deals with the central phase of the integration concept. The BAP and the QCAP are merged into a single optimization problem to decide on berthing positions, berthing times, and the assignment of cranes to vessels. The resulting problem is named the Berth Allocation and Crane Assignment Problem (BACAP). A mathematical formulation for this problem is provided where a major concern is to model crane productivity losses caused by crane interference and by berthing vessels apart from desired berthing positions. Several heuristic solution methods are described and intensively tested.

Precise handling times of vessels are derived from solving a QCSP. The high relevance of precise vessel handling time information for seaside operations planning necessitates a detailed study of the QCSP in Chap. 7. A mathematical formulation from the literature is taken up and revised by stating a new set of constraints for avoiding interference conflicts between cranes. A heuristic solution method is proposed for solving the problem. Afterwards, the QCSP formulation is extended to respect time windows for the cranes, and the heuristic is adapted to this problem variant. Again, computational tests are carried out to investigate the performance of the heuristics.

A final study in Chap. 8 deals with the integration of quay crane scheduling into the BACAP. This integration takes place in the first and the last phase of the new designed three-phase integration concept. In the first phase, a preprocessing of crane schedules is used to provide precise crane productivity information to the BACAP. This information supports the assignment of appropriate crane capacity to vessels. In the last phase, crane schedules are determined for the vessels to complete the outcome of the seaside planning process. Inconsistencies between a berth plan and the crane schedules are resolved by a repair procedure. The effectiveness of the two integration phases is finally assessed by computational tests.

Chapter 9 concludes the thesis. It summarizes the findings and outlines directions for further research.

Chapter 2
Maritime Container Transport

This chapter provides an introduction to the maritime container transport industry. Section 2.1 briefly describes the development of maritime container transport and the observed trends within the last decades. The organization of container transports is explained in sect. 2.2. In sect. 2.3 the layout of a container terminal and the available equipment are described.

2.1 A Brief History

In 1956 a freight forwarder named Malcolm McLean transported 58 specially designed containers on the vessel *Ideal-X*. This event is commonly seen as the birth of the civil maritime container transport industry. While earlier attempts at containerization have been made in the military and in the civil sector, McLean's achievement is the first implementation of a whole transport system completely aligned to the purpose of fast container transport and handling, see Levinson (2006). A substantial decrease in the handling time of the vessel and in the amount of laborers required for the transshipment process proved his concept to be far more profitable than conventional cargo handling. Soon regular liner services were established. The first services connected ports of the US east coast with ports in the Caribbean and in Central America. Later, services where established connecting ports all around the world. The port of Hamburg (Germany), for example, served its first container vessel in 1967, see HHLA (2008).

To control the development of different container systems, an international standardization of container measures was achieved in 1964, yielding a set of container sizes that were to be used from there on. The basic container unit today is of size $20' \times 8' \times 8'6''$ (length × width × height), also referred to as a TEU (Twenty-foot equivalent unit). The containers prevailing in maritime, road, and rail transport have a length of $40'$ feet, represented by two TEU, but also referred to as FEU (Forty-foot equivalent unit). Special purpose containers, such as "High Cube" containers

F. Meisel, *Seaside Operations Planning in Container Terminals*,
Contributions to Management Science,
DOI: 10.1007/978-3-7908-2191-8_2, © Physica-Verlag Berlin Heidelberg 2009

for cargo that overshoots a height of 8′6″, can differ in size. The container used in maritime, road, and rail transport can, however, not be used in air transport. The air transport industry has developed specialized container systems called ULDs (Unit Load Devices), adapted to the needs of aircraft, see IATA (2007).

Several trends can be identified in the maritime container transport industry. Within a few decades, containerization of general cargo became predominant. For example, in 2006 the degree of containerization was 97.2% in the Port of Hamburg, see Fig. 2.1. This change was accompanied by a growth of the world's container vessel fleet in terms of the number of vessels as well as in the size of vessels. In recent years this growth still continues due to the strong increase in international trade. For example, the number of container vessels with a gross tonnage of at least 300 tons grew from less than 2,500 in the year 2000 to more than 4,200 vessels in the year 2008, see ISL (2008). At the same time the total transport capacity grew from 4.4 million TEU to about 11 million TEU. According to their size, container vessels are classified into so-called generations. Although the vessel size of a particular generation is not standardized, approximate dimensions are shown in Fig. 2.2. The largest vessels in use today have a capacity of about 11,000 TEU.

The trend of increasing vessel size still continues but the application of so-called Ultra-Large Container Ships (ULCSs) seems to be limited for several reasons. First, a proper travel speed of a ULCS requires major constructional changes, e.g., the

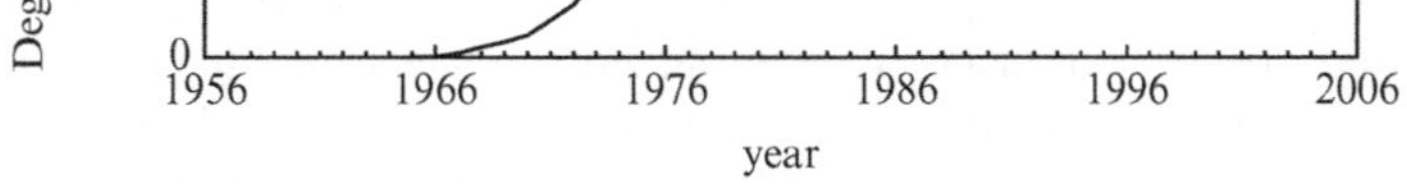

Fig. 2.1 Degree of containerization in the port of Hamburg (*data source:* Port of Hamburg Marketing, 2008)

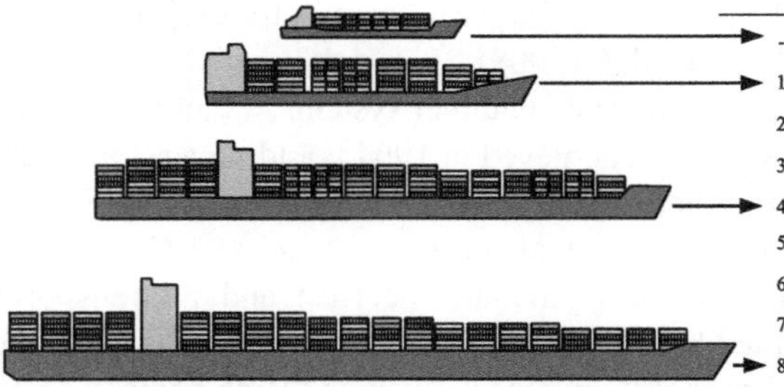

Generation	Capacity [in TEU]	Length [in m]	Beam [in m]	Draft [in m]
-	< 760	< 120	< 16	< 8
1.	760-1000	120-190	16.0-28.0	8.0-10.0
2.	2000-2800	210-240	28.0	11.5
3.	3000-4000	260-290	32.2	12.5
4.	4000-5000	280-295	32.2	13.5
5.	5000-6000	285-318	39.2-40.8	13.5
6.	6000-6400	295-318	40.0-42.8	14.2
7.	6400-7500	318-348	42.8-45.0	14.8-15.0
8.	7500-8400	348-365	~48.0	14.8-15.2

Fig. 2.2 Vessel size and capacity by generations (*data source:* Brinkmann, 2005, p. 67)

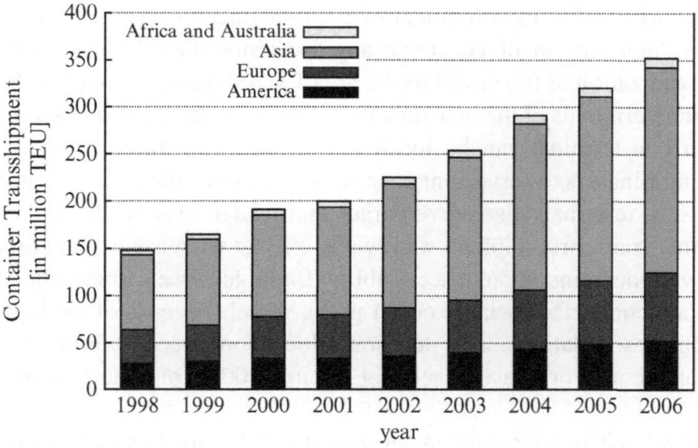

Fig. 2.3 Container transshipment by continents (*data source:* Port of Hamburg Marketing, 2008)

installation of a second engine, which result in a considerable jump in the construction cost, see Tozer and Penfold (2001). A decrease in travel speed is unacceptable because the advantage of the capacity increase may be canceled out by a reduced number of trips per year. Second, the larger the vessel is, the smaller is the number of ports and canals it can pass through. Such restrictions already exist for vessels with a beam exceeding 32.2 m. They are unable to pass the Panama Canal and are therefore also named Post-Panamax vessels, see Brinkmann (2005). Third, the number of routes where ULCS can be profitably applied is further limited to those that show a sufficiently large transport demand. According to Tozer and Penfold (2001), a most promising trade route for the application of ULCS is the one between East Asia and Europe.

A further trend in maritime container transport is the high growth in the volume of transshipped containers in ports. Figure 2.3 shows the container transshipment by continents within the years 1998–2006. It can be seen that the total number of transshipped container units has more than doubled within this time span, showing a total of 350 million TEU in 2006. This growth reflects the increasing transshipment demand caused by the attractiveness of containerized cargo transport.

2.2 Organization of Container Transports

Container transport takes place within road, rail, inland waterway, and maritime traffic networks. A transshipment node in a container transport network is referred to as a *container terminal* (CT). Usually traffic networks overlay with each other and terminals can be part of more than one network. In such terminals, containers can be transshipped between different modes of transport. Forwarding a container from a shipper to a recipient requires the use of transport relations of one or more traffic networks and a transshipment of the container in a CT whenever different transport vehicles are involved. Regarding this issue, the role of seaport terminals and the organization of the vessel traffic are of particular relevance for this thesis.

Container terminals of the maritime traffic network are located in seaports, where more than one terminal can be located in a port. The main purpose of seaport container terminals is to serve container vessels. Besides the large ocean-going container vessels, terminals also serve barges and feeder vessels. Barges are used for the container transport on inland waterways. Feeder vessels connect ports with low transport volume or insufficient accessibility for large vessels to so-called hub ports. The hubs are connected by large ocean-going vessels because of the high transport volume on these relations. The decision which ports become hubs are a subject of the strategic network design, see, e.g., Baird (2006) for a study concerning this matter.

Serving vessels in a terminal means loading and unloading containers. Containers to load on a vessel are commonly referred to as *export containers*. Containers to unload from a vessel are referred to as *import containers*. The export containers are delivered from the shippers to the terminal ahead of the sea trip. The import containers have to be delivered to the recipients after their arrival at the terminal. These two types of transports constitute the pre- and post-carriage for the transport by an ocean-going vessel. The pre- and post-carriage can be executed using different combinations of transport means, as there are trucks, trains, barges, or feeder vessels, see Fig. 2.4. Since most shippers and recipients are not directly connected to a railway network or to a port, truck transport is typically involved in a practical transport chain. Next to container transshipment, terminals also intermediately store containers. This allows delivery of export containers from the hinterland before the arrival of the dedicated vessel and pickup of import containers for further transport after the delivering vessel has already departed. In other words, a temporal decoupling of the transport links is enabled.

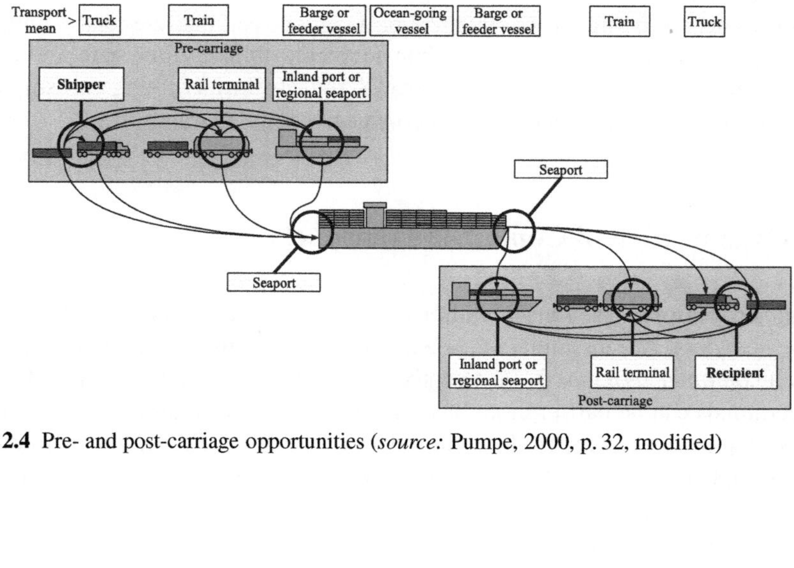

Fig. 2.4 Pre- and post-carriage opportunities (*source:* Pumpe, 2000, p. 32, modified)

The region made up of the origins of export containers and of the destinations of import containers is called the *hinterland* of a terminal. Since terminals closely located to each other show different reachability in terms of transport distances, transport times, and transport connections, the hinterland of a specific terminal cannot be defined exactly and may reach into the hinterland of neighboring terminals, see Lemper (1996). In such a situation, terminals compete with each other in attracting container shipments. An example constitutes the so-called North Range in Europe, where major CTs are located in Le Havre, Rotterdam, Antwerp, Bremerhaven, and Hamburg.

For connecting the seaport terminals, *liner services* are established on which container vessels operate. In a liner service, vessels follow a fixed schedule that gives the order of ports to visit and the calling times, see Ronen (1983). For a liner service it has to be decided on the ports to connect and on the size of the vessels to deploy. Furthermore, by deciding on the frequency of visiting a port within a schedule (e.g., on a weekly call basis) the number of vessels to deploy is determined. Liner services can be performed in round-the-world tours, where the sequence of ports lets the vessels go around the globe, or on a pendulum tour between two or three important trade regions. On a pendulum tour a vessel visits its ports in a given order and, at the last port, reverses this order to end up in the port where the tour began.

The described organization of transport networks and transport operations coincides with the (economic) advantages that make the transport of containerized goods so attractive. The shippers take advantage of the regular transport opportunity offered by the liner service schedules. The various ways of realizing pre- and post-carriage enable shippers to flexibly align transport operations to their individual requirements. A further important aspect is that vessel capacity can be booked on the basis of single TEU, providing a flexible transport opportunity almost independent of the actual volume of a shipment. A large variety of container types enables shippers to use container transport not only for general cargo but also for liquid, bulk, frozen, and other types of goods, see Hapag-Lloyd (2008). Finally, the fast transshipments decrease the overall transport time and thus, reduce the shippers' cost of capital tied up in the transported goods.

The vessel operators benefit from *economies of scale*, which motivate to establish hub terminals in order to utilize large-sized container vessels. Port related and travel related economies of scale are observed. The first stems from the fact that a relative increase in the size of a vessel leads to an underproportional increase in its port cost. For example, according to Stopford (1997), port cost per call (without cargo handling cost) has been 22,000 USD for a 1,200 TEU vessel but only 43,000 USD for a 6,500 TEU vessel in the year 1996. The economies of scale related to the travel of a vessel are based on similar observations regarding an underproportional increase in operational cost, capital cost, and bunker cost per day, leading to costs per TEU per day of 16.6 USD for a 1,200 TEU vessel and 7.5 USD for a 6,500 TEU vessel, see Stopford (1997).

Containerization also leads to *economies of speed*. From the point of view of vessel operators only the travel time of a vessel is economically productive. The speedup of cargo transshipment as enabled by containerization reduces the handling times of vessels in ports from days or even weeks down to hours. Reductions in vessel handling times increase the proportion of travel time allowing for more trips per year and the generation of revenue for additionally transported containers. This revenue can basically be obtained by an increase in the travel speed, too. However, the speed up of cargo handling seems to be more profitable compared to the increase in travel speed, see Laine and Vepslinen (1994). Consequently, the performance of terminal operations is crucial for the profitability of liner services.

2.3 Layout and Technical Equipment of a Container Terminal

The layout of a seaport container terminal consists of different areas each one serving a specific functional purpose. The four major area types are:

- *Quay area* for mooring the container vessels
- *Transport area* for the transport of containers within the terminal
- *Yard area* for the storage of containers
- *Truck and train area* for serving the external trucks and the trains

Various technical equipment is used for the terminal operations. Cranes are employed at the quay and in the yard. Yard trucks or Automated Guided Vehicles (AGVs) perform the transport of containers between the terminal areas. Alternatively, so-called straddle carriers or Automated Lifting Vehicles (ALVs) can perform the transport as well as the stacking operations in the yard. Figure 2.5 sketches the functional areas and the equipment alternatives. In the following, a brief description of the areas and the equipment is provided. More detailed descriptions are given by Brinkmann (2005).

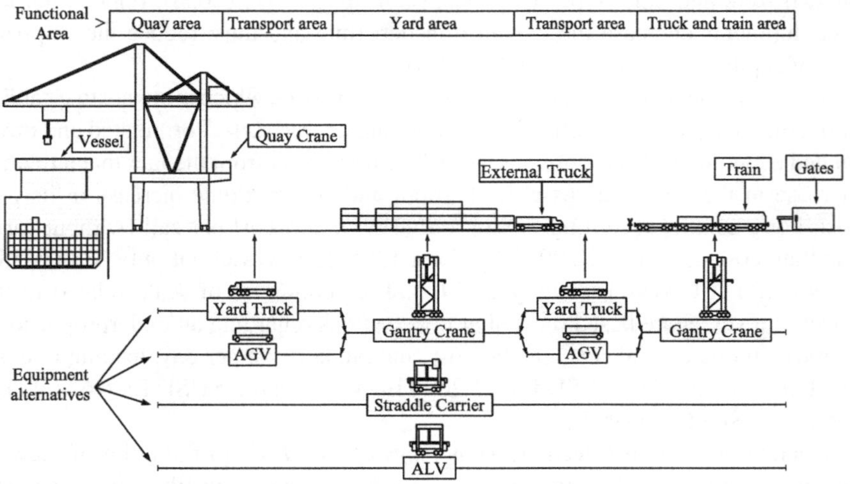

Fig. 2.5 Schematic cross sectional view of a container terminal

2.3.1 Quay Area and Quay Cranes

The seaside functional area of a CT is the quay, where ocean-going vessels, feeder vessels, and barges moor. The loading and unloading operations of containers are performed by *quay cranes* (QCs), see Fig. 2.6. Some container vessels are self-equipped with cranes in order to enable transshipment operations independent of the equipment offered at a terminal. However, nowadays vessel operators usually abstain from this option because a sufficient standard of equipment is offered at most terminals. Depending on their size, vessels may be served by up to six QCs simultaneously. Large vessels can have up to 22 container stacks side by side in a bay requiring properly dimensioned cranes with an outreach of 60 m, see Tozer and Penfold (2001). Due to the difficulties of accessing the containers within a vessel, a skilled crane driver is needed to operate a QC.

A loading or unloading operation of a container is referred to as a *move*. To unload a container, the QC's spreader is placed on it, fixed by twistlocks, and then lifted by a hoist. The crane's trolley moves to the quay where the spreader is lowered and the container is either put on the ground or put on a transport vehicle. The container is released by unlocking the twistlocks, and the spreader is hoisted again. The loading of a container uses the same crane operations in the opposite direction.

The productivity of a QC is measured by the number of moves per hour. This is a key indicator for the productivity of a terminal. In practice a QC currently realizes about 30 moves per hour, see Chu and Huang (2002). However, technological

Fig. 2.6 Quay cranes serving a container vessel (*source:* Port of Hamburg Marketing / D. Hasenpusch, 2008)

improvements aim at increasing crane productivity. Vessels can be equipped with cell guides to ease the positioning of containers within the hold, see Goedhart (2002). QCs can be equipped with two trolleys. One trolley serves the vessel and the other trolley serves the horizontal transport vehicles. Containers are handed over from one trolley to the other at a platform between the crane's uprights. While the landside trolley can be automated, an operator is still needed for the seaside trolley, see Jordan (2002). These so-called Dual Trolley QCs are applied, for example, in the Container Terminal Altenwerder in Hamburg.

2.3.2 Transport Area and Transport Vehicles

The horizontal transport system of a CT moves containers between the functional areas. The most often used equipment types are yard trucks and straddle carriers. Yard trucks are manned vehicles that pull chassis carrying the containers, see Fig. 2.7a. They are unable to lift containers and thus, demand a crane for loading and unloading operations. This requires a careful synchronization of cranes and trucks to avoid crane waiting. Dual Trolley QCs can be used to reduce such idle times because they enable a temporal decoupling of (un-)loading operations of the vessel and of the trucks. Yard trucks represent the technologically modest way of the container transport. Nevertheless, this can be economically attractive because of low purchase and maintenance costs as well as high flexibility regarding the workload of a terminal. However, labor costs for drivers lead to high operational costs.

An alternative to the use of yard trucks are straddle carriers, see Fig. 2.7b, also referred to as van carriers. Next to moving containers they are also able to lift and stack containers. This allows for the decoupling of QC operations and transport operations. If straddle carriers are used, a QC can put an unloaded container on the quay and continue the service process. This avoids crane waiting and leads to increased crane productivity in terms of moves per hour. The only prerequisite for unloading operations without crane waiting is that a free ground position is available to drop a container. Loading operations without crane waiting are enabled if straddle

(a) (b)

Fig. 2.7 Yard truck (**a**) and straddle carrier (**b**) (*source:* Kalmar Industries, 2008)

(a) (b)

Fig. 2.8 AGV at the port of Hamburg (**a**) and ALVs at the port of Brisbane (**b**) (*source:* Gottwald Port Technology, 2008; Kalmar Industries, 2008)

carriers deliver containers before the QC is ready for the pick up. The carriers can move away after a delivery without having to wait for the QC. Compared to yard trucks higher purchase, maintenance, and operational costs come along with the employment of straddle carriers.

Yard trucks and straddle carriers can be replaced by fully automated alternatives, namely Automated Guided Vehicles (AGVs) and Automated Lifting Vehicles (ALVs), see Fig. 2.8. The movement of automated vehicles is guided by induction coils installed in the pavement. AGVs are able to carry one 40′ container or two 20′ containers. ALVs transport a single container but can lift it like straddle carriers. Automated transport systems in CTs do currently not reach high transport productivity. One reason is that automated vehicles usually show a comparable low average speed. Another reason is that a breakdown of an automated vehicle may lead to downtime of the entire transport system. Nevertheless, the promised advantages of automated vehicles are the reduction of labor cost at the terminal and the reliable execution of work plans resulting from the elimination of human failure. Since the investment in automation needs to pay off, it is especially attractive for terminals with a high labor cost level, see Nam and Ha (2001). Currently AGVs are employed at Container Terminal Altenwerder (Hamburg, Germany), at European Combined Terminals (Rotterdam, Netherlands), and at Pusan Eastern Container Terminal (Pusan, South Korea). ALVs are a relatively new development used at the CT of Brisbane (Australia).

The functional area for executing horizontal transport is typically divided into one or more traffic lanes. The seaside transport area, also referred to as the apron, usually contains three traffic lanes, two for the flowing traffic and one for vehicles waiting at QCs to be served. The width of the traffic lanes depends on the equipment used. Yard trucks pulling several chassis at a time require a larger turn radius compared to straddle carriers. AGVs and ALVs require less traffic space because of the low speed and the precise guidance. While manned vehicles can flexibly use the available traffic lanes, the movement of automated vehicles is restricted to the network of designated travel routes. In the simplest case a single unidirectional loop design is used. Here, vehicles move along the quay in one direction and along the yard in the other direction. The vehicles can temporarily exit the loop to pickup and deliver containers at a vessel or at the yard. More complex network designs allow, for example, for bidirectional usage of traffic lanes or for more flexible routings of vehicles in order to shorten the travel distances, see Schrecker (2000).

2.3.3 Yard Area and Yard Cranes

The yard is used for the intermediate storage of containers. Import containers are stored until the hinterland transport is initiated. Export containers are stored until they are loaded onto the dedicated vessels. Areas for the storage of empty and reefer containers exist as well. A yard is usually divided into a set of yard blocks, which are separated by traffic lanes. A block consists of several parallel container rows, each of them providing a number of lengthwise arranged storage positions, see Fig. 2.9a. Multiple tiers of containers can be stacked at each position.

In Fig. 2.9a a yard is shown where the stacking and retrieval operations are performed by gantry cranes. These cranes can be rail mounted gantry cranes (RMGCs) or rubber tired gantry cranes (RTGCs). Depending on its design a RMGC spans up to 13 container rows, see Linn et al. (2003). A RTGC spans up to 8 container rows only, but it can be repositioned to other yard blocks, see Kalmar Industries (2008). Both crane types stack containers up to six tiers high. Usually the upmost tier of each stack remains empty in order to allow a crane passing over the stack with a container. One of the rows may be reserved for the service of transport vehicles, see Fig. 2.9a. Alternatively, vehicles are served at a front side of a block, which allows for a higher storage capacity but requires additional movement of the gantry cranes. Advanced technological solutions increase the transshipment capability of a crane operated yard. Double Rail Mounted Gantry Cranes possess two cranes of different size operating within the same block. The different sizes enable the cranes to pass each other, allowing for more flexible operations. Gantry cranes can be automated to a large extent. Such automated Double Rail Mounted Gantry Cranes are used at the CT Altenwerder in Hamburg. If straddle carriers are used for yard operations, the block structure is broken up by additional clearance between the container rows, see Fig. 2.9b. This enables straddle carriers to enter each row and access the desired

Fig. 2.9 Container storage within the yard (*source:* Kalmar Industries, 2008)

storage position. Straddle carriers can stack containers up to a height of four tiers. Again, the upmost tier usually remains empty. If yard trucks are used at the CT, containers can be stored on the transport chassis in the yard. Obviously, little investment and operational costs as well as direct access to every container is possible, but storage capacity is wasted at the same time. Storage capacity is a key performance indicator for terminals where storage space is scarce. In general, the best storage capacity is achieved if gantry cranes are used. The usage of straddle carriers leads to a lower storage capacity which, however, is still much higher than the capacity enabled by storing containers on chassis.

2.3.4 Truck and Train Area

A seaport terminal provides the interface to the hinterland by serving trains and external trucks. Trucks have to pass gate houses, where containers are checked and transport documents are processed. If straddle carriers are used for the yard operations, trucks move to a parking area in order to be served. If gantry cranes are used, trucks are sent directly to the dedicated yard blocks in order to be served. Self-service of trucks is possible if containers are stored on chassis in the yard. For the service of trains, railway tracks lead into the terminal. If the yard is operated by gantry cranes, trains are served by gantry cranes too, where horizontal transport of containers is required again. Otherwise, straddle carriers or ALVs are used.

As can be seen above, a CT can be operated either by straddle carriers (ALVs) alone or by a combination of gantry cranes and yard trucks (AGVs). Since every equipment type shows its own strengths and weaknesses, there exists no overall best equipment selection. The selection decision basically aims at a high transshipment capability and at an economic balance of the investments to make and the expected operational costs. But also local conditions, such as a limitation of space or the labor education level, may enforce a certain decision. The particular equipment selection that is implemented in a terminal constitutes a set of requirements for the management of terminal operations.

Chapter 3
Operational Planning Problems

In this chapter the operational planning problems of a CT are informally described. In Sect. 3.1 a distinction of planning levels is discussed. The planning of seaside operations, internal operations, landside operations, and workforce utilization are described in Sects. 3.2, 3.3, 3.4 and 3.5 respectively.

3.1 Distinction of Planning Levels

The planning problems of a CT can be distinguished, depending on their planning horizon, into strategic, tactical, and operational planning problems. Strategic decisions concern the location and layout of a newly built terminal as well as decisions regarding the types and number of equipment to use. The degree of automation of the CT is thereby defined. Strategic decisions usually last for years. At the tactical level it is decided on the space usage of the terminal, for example, by determining the yard blocks for the storage of empty and reefer containers and the layout of traffic courses for the horizontal transport system. These decisions last for months, or at least for weeks. At the operational level, work plans for the CT resources are generated in order to realize the service of vessels, trucks, and trains. These decisions last for several days down to seconds only.

An alternative problem classification is provided by Günther and Kim (2005) who distinguish *Design Problems*, *Operational Planning Problems*, and *Real Time Control Problems*. While the first can be seen as an aggregation of strategic and tactical decisions, the latter refers to decisions which need to be made while service processes are executed. Such problems, e.g., the assignment of a transport order to a vehicle, have to be solved in accordance with the current state of the CT system. Real time control problems are characterized by a very short planning horizon of a few minutes and by the need to generate solutions very quickly, usually within seconds. A further classification of CT planning problems is provided by Lehmann (2006) who refers to strategic and tactical problems as *Terminal Design* and to operational

F. Meisel, *Seaside Operations Planning in Container Terminals*, Contributions to Management Science,
DOI: 10.1007/978-3-7908-2191-8_3, © Physica-Verlag Berlin Heidelberg 2009

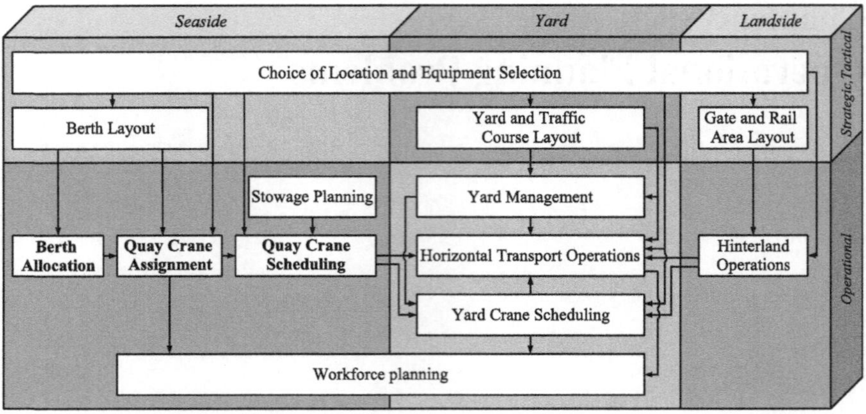

Fig. 3.1 Planning problems in container terminals

problems as *Terminal Logistic*. Figure 3.1 provides an overview of the corresponding planning problems at the seaside, the yard, and the landside together with the basic problem interdependencies.

In the following the operational planning problems of a CT are described. Those problems dealing with berth planning, QC assignment, and QC scheduling are described in higher detail because they are in the focus of this thesis. Literature surveys follow for these problems in the next chapter. Other operational planning problems shown in Figure3.1 are only described briefly. For a comprehensive overview of CT planning problems and related literature surveys the reader is referred to Meersmans and Dekker (2001), Vis and de Koster (2003), Steenken et al. (2004), Vacca et al. (2007), and Stahlbock and Voß (2008).

3.2 Seaside Operations Planning

3.2.1 Berth Allocation

In the *Berth Allocation Problem* (BAP) we are given the berth layout of a CT together with a set of vessels that have to be served within the planning horizon. For each vessel additional data like the vessel's length including clearance, its draft, the expected time of arrival, and the projected handling time, i.e., the duration of the vessel's service, can be given. All vessels must be moored within the boundaries of the quay. They are not allowed to occupy the same quay space at a time. The problem is to assign a berthing position and a berthing time to each vessel, such that a given objective function is optimized. Berth planning has been shown to be an $\mathcal{NP}$-hard problem by relating it to the set partitioning problem (Lim, 1998), the

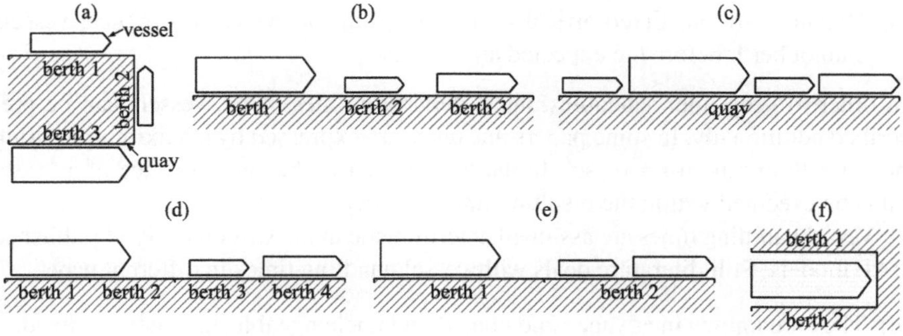

Fig. 3.2 Berth and quay relationship

single machine scheduling problem with release dates (Hansen and Oğuz, 2003), and the two dimensional cutting stock problem (Imai et al., 2005).

There may be further constraints involved in berth allocation, which leads to a multitude of BAP formulations. Spatial constraints restrict the feasible berthing positions of vessels according to a preset partitioning of the quay into berths. According to Imai et al. (2005) the following cases are distinguished:

(a) Discrete layout: The quay is partitioned into a number of sections, called berths. Only one vessel can be served at each single berth at a time. The partitioning can either follow the construction of the quay (Fig. 3.2a) or is organizationally prescribed to ease the planning problem (Fig. 3.2b).
(b) Continuous layout: There is no partitioning of the quay, i.e., vessels can berth at arbitrary positions within the boundaries of the quay (Fig. 3.2c). For a continuous layout, berth planning is more complicated than for a discrete layout at the advantage of better utilizing quay space.
(c) Hybrid layout: Like in the discrete case, the quay is partitioned into berths, but large vessels may occupy more than one berth (Fig. 3.2d) while small vessels may share a berth (Fig. 3.2e). An indented berth results if two opposing berths exist, which can be used to serve a large vessel from both sides (Fig. 3.2f).

In case of draft restrictions further spatial constraints must ensure that vessels are berthed at positions of sufficient water depth. To avoid enlarging the handling time, container vessels usually stay at the assigned berthing position during the entire service. In contrast, vessels are allowed to be repositioned in naval ports and general cargo terminals, see, e.g., Brown et al. (1994, 1997), and Lee and Chen (2008).

Temporal constraints can restrict the berthing times and the departure times of vessels. According to Imai et al. (2001) the following cases are distinguished:

(a) Static arrival: There are no arrival times given for the vessels or arrival times impose merely a soft constraint on the berthing times. In the former case it is assumed that vessels already wait at the port and can berth immediately. In the latter case it is assumed that a vessel can be speeded up at a certain cost in order to meet a berthing time earlier than the expected arrival time.

(b) Dynamic arrival: Fixed arrival times are given for the vessels, hence, vessels cannot berth before the expected arrival time.

In order to keep liner schedules, latest departure times of the vessels can be prescribed additionally. In some papers due dates are expressed by a maximum waiting plus handling time for a vessel. In the dynamic case, the entire service of a vessel must be executed within the resulting time window.

Vessel handling times are assumed deterministic in the vast majority of published BAP models. Still, literature deals with vessel handling times in different ways:

(a) They are known in advance and considered unchangeable, i.e., they are fixed.
(b) They depend on the vessels' berthing positions.
(c) They depend on the number of cranes serving the vessels.
(d) They depend on the work schedules of the assigned cranes.
(e) They obey to combinations of (b), (c), and (d).

The general goal of berth planning is to provide fast and reliable services of vessels. This is reflected in the literature by various objective functions. Models to minimize the sum of the waiting and handling times of vessels (i.e., the port stay times) clearly prevail. Further objectives are, for example, the minimization of the workload of terminal resources and the minimization of the number of vessels rejected to be served at a terminal. The performance of a berth plan is often measured in terms of costs which allows to combine different goals in an overall cost function.

A solution to the BAP, the so-called berth plan, is typically represented in a space–time diagram as shown in Fig. 3.3. In this example the quay is not partitioned into berths. The scaling used for the space–time diagram divides the spatial axis into quay segments of 10 m length, making up a total quay length of 600 m in the example, and the time axis into periods of 1 h. The service process of each vessel is represented by a rectangle. The height of a rectangle corresponds to the length of the vessel (including clearance) and the width corresponds to the expected handling time. The lower-left vertex of a rectangle gives the berthing time and the

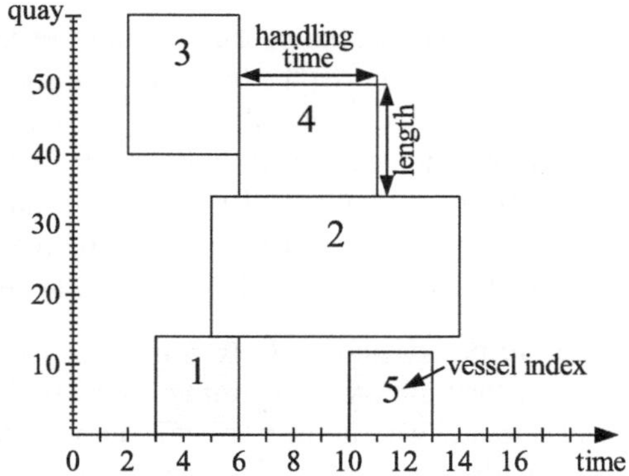

Fig. 3.3 Space–time representation of a berth plan

berthing position, e.g., Vessel 3 berths at time 2 at position 40. Since no rectangles overlap and all vessels are moored within the boundaries of the quay, this solution is feasible, assumed that no further restrictions are given.

3.2.2 Quay Crane Assignment

In the *Quay Crane Assignment Problem* (QCAP) we are given a feasible berth plan and a set of identical QCs, which are available for service. For each vessel included in the berth plan, the volume of containers to be loaded and unloaded is known as well as the maximum number of cranes allowed to serve it simultaneously. The cranes are supposed to be lined up alongside the quay. They can be moved to every vessel but they are not able to pass each other. The problem is to assign cranes to vessels such that all required transshipments of containers can be fulfilled. The need to solve the QCAP arises from the fact that, beside the quay space, also the QCs are a scarce resource at a CT. Due to high purchase and maintenance costs, terminal operators usually provide less QC capacity than the technical possible limit. This allows for a high utilization rate of the available cranes but, obviously, can affect a bottleneck if multiple vessels are served simultaneously. The distribution of cranes to vessels is sometimes referred to as the crane split, cf. Steenken et al. (2004). QCAP and BAP are basically interrelated, because solving the QCAP can have a strong impact on the vessels' handling times. Only in case of a discrete berth layout, where each berth holds a set of dedicated cranes, an explicit assignment of cranes to vessels is not necessary.

The decisions of the QCAP can be broken down into:

1. Determine the number of QCs to assign to each vessel.
2. Determine the specific QCs that make up the set of assigned cranes.

For the first decision, several basic constraints exist. Each vessel has to receive sufficient crane capacity for its service. In the assignment of crane capacity it should be taken into account that additional cranes assigned to a vessel show a decreasing marginal productivity due to crane interference problems. The assignment of cranes to vessels has to respect also the berthing times as given by a preset berth plan. Obviously, the available number of QCs at a terminal must be respected at all times. An additional constraint of the QCAP is that the number of cranes serving a vessel simultaneously is often restricted. A minimum number of cranes can be contracted between the vessel operator and the CT operator while a maximum number defines a technical limit. If the minimum number of cranes is larger or equal to one, a vessel is served without preemption. Chu and Huang (2002) provide empirical data for different terminals of the port of Kaohsiung (Taiwan), which reveals that the majority of vessels is served by two to three QCs while large vessels are served by up to six cranes in parallel.

Figure 3.4a shows one crane capacity assignment for the berth plan depicted in Fig. 3.3 with four cranes available at the quay. A gray shaded rectangle within

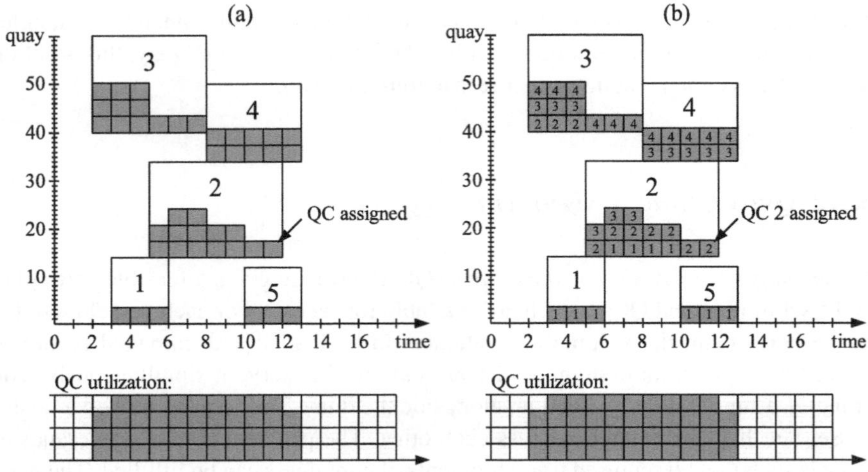

Fig. 3.4 Assignment of crane capacity (**a**) and assignment of specific cranes (**b**)

a vessel-rectangle indicates the assignment of one QC in the corresponding time period. As can be seen in this example, assigning more cranes to a vessel can accelerate its handling time at the expense of other vessels. For example, the finishing of Vessel 2's service is projected at time 14 in Fig. 3.3 but it is already completed at time 12 under the crane assignment shown in Fig. 3.4a. The finishing time of Vessel 3, however, increases from time 6 to time 8 as a result of the crane assignment. If handling times increase, rectangles may overlap in the space–time representation. In this case, the berth plan is no longer feasible. A revision of the BAP decisions is required, as done in the berth plan of Fig. 3.4a where the berthing time of Vessel 4 has been delayed. The QCAP can also identify improvement potentials if vessels can depart earlier than projected in the berth plan. Hence, the minimization of the additional delays or the maximization of early departures are important objectives of the QCAP.

The assignment of cranes to a single vessel is called a *QC-to-Vessel assignment*. One can distinguish between time-invariant and variable-in-time assignments. In a *time-invariant* QC-to-Vessel assignment the number of cranes assigned to a vessel is constant during the complete handling time (see Vessels 1, 4, and 5 in the example in Fig. 3.4). In a *variable-in-time* QC-to-Vessel assignment the number of cranes assigned to a vessel can change during the handling time (Vessels 2 and 3 in Fig. 3.4). Variable-in-time assignments are very common in practice because they enable a flexible assignment of crane capacity to vessels. Chu and Huang (2002) report, for the port of Kaohsiung (Taiwan), that the crane ready times differ by more than 1 h for 50% of the large vessels, i.e., variable-in-time QC-to-Vessel assignments are realized. Another advantage of a variable-in-time assignment is that cranes can be removed from a vessel before the service is completed, in order to serve another vessel of higher priority. Of course, under time-invariant crane assignments the QCAP is easier to solve, but the service quality provided to the vessels and the utilization of QCs are lowered in general.

The second decision of the QCAP is to determine the *specific* QCs that make up the set of assigned cranes. Since QCs are mounted on a rail track alongside the quay, they cannot cross (pass) each other. Therefore, a specific assignment is only feasible if the relative positions of cranes are preserved at all times. Figure 3.4b shows an assignment of specific QCs, where the cranes are indexed according to their relative positioning at the quay. Since the cranes do not cross, the assignment is feasible. The specific QC-to-Vessel assignments determine the number of crane setups at vessels and the imposed movement of cranes between vessels. Hence, minimizing the number of crane setups or the crane travel times are important objectives when deciding on the specific cranes that are used for the service of vessels.

In practice, the QCAP is not found a difficult problem if solved by rules of thumb. Therefore, the problem has hardly received attention by its own in academic research. Due to the profound impact on vessels' handling times, however, crane assignment decisions are involved in some advanced berth planning models.

3.2.3 Quay Crane Scheduling

In the *Quay Crane Scheduling Problem* (QCSP) we consider a set of tasks, representing transshipment operations for a vessel, and a set of assigned QCs. Precedence relations among tasks can be given to ensure that unloading precedes loading and to represent the stacking of containers as defined by a stowage plan. Every task must be processed (usually without preemption) once by a QC while a QC can process at most one task at a time. A solution to the problem, called a QC schedule, defines a starting time for every task on a crane. Usually, the minimization of the makespan of the QC schedule is pursued because it represents the handling time of the considered vessel. The problem described thus far corresponds to a minimum makespan scheduling problem with parallel identical machines and precedence constraints. This problem is known to be $\mathcal{NP}$-hard in the strong sense, provided that more than two machines (cranes), non-preemption or non-uniform processing times are given (Pinedo, 2002).

To avoid crossing of cranes, the QCSP requires a spatial constraint, which is not involved in machine scheduling problems. As a further spatial constraint, sophisticated QCSP models also comprise the compliance with safety margins between adjacent cranes. Non-crossing and safety margins can prevent cranes from reaching their maximum transshipment productivity. In general, the more QCs are assigned to a vessel, the more crane interference will be observed. This leads to the phenomenon of a decreasing marginal productivity. Additional attributes for tasks and cranes lead in fact to a variety of different models for QC scheduling.

Tasks to be scheduled on a QC describe the granularity in which the workload of a vessel is considered in a QCSP model. Tasks can be defined on the basis of bay areas or single bays (Fig. 3.5a), or on the basis of container stacks, container groups, or individual containers (Fig. 3.5b):

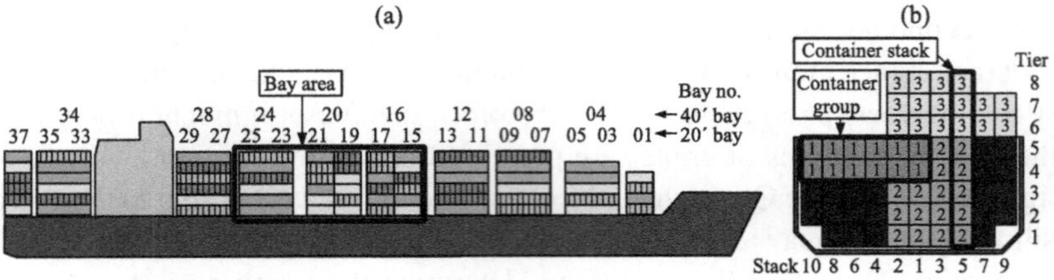

Fig. 3.5 Storage location structure of a vessel (**a**) and a bay (**b**)

(a) Bay areas: A task consists of all loading and unloading operations of containers within a certain area of the bays.
(b) Bays: A task consists of all loading and unloading operations in a bay.
(c) Stacks: A task consists of the transshipment of all containers in a stack.
(d) Groups: A task refers to a group of containers that are stored in adjacent slots of a bay. Grouped containers usually have a common destination, or the like.
(e) Containers: A task consists of the loading or unloading of a single container.

The idea of dividing the workload of a vessel into bay areas is to serve each bay area exclusively by one QC. If the bay areas are non-overlapping, crane interference is completely avoided. However, a sufficient balance of the workload distribution among the cranes is often not possible. In case (b), a better workload distribution might be achievable, but solving the problem becomes much more complicated due to the required non-crossing constraint. As the number of tasks is bounded by the size of the vessel, the problem complexity is still moderate. Reducing the granularity further on allows to improve crane schedules at the expense of increasing the problem complexity. With hundreds to thousands of tasks, as observed for a large vessel in case (e), the corresponding QCSP might become intractable.

Next to task attributes, also crane attributes appear in QCSP models to specify the crane operations in more detail:

(a) Ready times: An individual ready time is used to designate the earliest possible operation of a crane.
(b) Time windows: For each crane one or more time windows can be used to specify time spans where the crane is available to serve the considered vessel. These time windows are often an outcome of variable-in-time QC-to-Vessel assignments.
(c) Positions: Initial and final positions are prescribed for a crane.
(d) Travel times: The speed of the crane movement is given in terms of the time required to travel between bays.

3.2.4 Stowage Planning

Stowage planning concerns the assignment of export containers to empty slots within a vessel. A slot is a container position in a bay, defined by a stack and a tier number as shown in Fig. 3.5b. Stowage planning is made on two levels of aggregation. First, a coarse stowage plan is generated by assigning container classes to slots. Second, a precise stowage plan is generated by assigning individual containers to slots of their container class. Wilson and Roach (2000) describe both levels in detail. In practice, the vessel operators determine the coarse stowage plan because they possess the information regarding the sequence of ports to call at and the expected number of containers per class to load and unload in each port. A basic constraint at this stage of the planning process is that the derived stowage plan has to ensure the stability of the vessel on its journey. Furthermore, the assignment of container classes to slots can be restricted. E.g., reefer containers can only be assigned to slots providing electric supply. One important objective is the maximization of the vessel's capacity utilization. A further objective is the minimization of the number of *reshuffles*. A reshuffle (also called a re-handle) is a temporal removal of a container which is not dedicated to the considered terminal but stays on top of containers to unload. Since reshuffles are unproductive container moves, minimizing their number leads to a short vessel handling time.

The stowage plan is specified when the vessel is served at a terminal. Export containers are assigned to slots with respect to the container classes as set by the vessel operator. The objective is again the minimization of reshuffles within the vessel but also the minimization of reshuffles within the yard. A good stowage plan always allows one to pick the upmost container from a yard stack while loading the vessel.

3.3 Internal Operations Planning

3.3.1 Yard Management

Yard management comprises three tasks:

- Reservation of yard capacity for liner services
- Selection of storage locations for individual containers
- Repositioning of containers within the yard (remarshalling)

With the first decision, complete or partial yard blocks are assigned to calling vessels for buffering the import and export containers. In order to smooth the workload among the gantry cranes, several yard areas can be reserved for a vessel which are not necessarily located close to each other. The reserved capacity has to fit to the expected transshipment volume of a vessel. The reserved space for export containers can be partitioned by container classes to avoid reshuffles while vessels are served. The objective of the yard reservation is to realize a high space utilization over time. Therefore, reserved space does not have to be available at the time of the reservation, but it has to become available before the containers arrive at the yard, see, e.g., Fu et al. (2007).

The selection of a storage location for an individual container requires to select a yard block first. The yard block for an export container is selected among the reserved yard capacity for the liner service. The selection of a storage location within a block pursues the minimization of reshuffles again. Dekker et al. (2006) show that a stacking policy, where export containers of a same class are stored in the same yard stack, reduces reshuffles at the time of the vessel service. The reduction is based on the fact that containers of the same class are exchangeable. Thus, the top container of a stack can always serve as the next container to be loaded. This strategy, however, yields no advantage for import containers. Import containers are not exchangeable and the pickup times for the hinterland transport is usually unknown. As a consequence, the determination of a stacking order minimizing the number of reshuffles is not possible. Nevertheless, reshuffles can be reduced by reducing the stacking height.

Yard management decides furthermore on repositionings of containers in the yard, also referred to as yard remarshalling. Repositionings become necessary if containers are not stored within the reserved areas or if their stacking order requires reshuffles while the service of a vessel takes place. The repositionings are performed in periods of low workload. According to Lee and Hsu (2007), remarshalling should aim at the minimization of container moves necessary to resolve inappropriate storage locations and stacking orders of containers.

3.3.2 Yard Crane Scheduling

The gantry cranes that operate within a yard are also referred to as yard cranes (YCs). Their scheduling comprises two tasks:

- Deployment of cranes to yard blocks
- Scheduling of stacking and retrieval operations for single containers

The first planning problem arises if the available number of YCs is lower than the number of blocks within the yard. In this case, cranes need to be moved to those blocks where stacking and retrieval operations have to be performed. The deployment of cranes is planned on a horizon of several hours. For technical reasons only one or two YCs can work within a block simultaneously. The objective pursued by the YC deployment is the minimization of an unfinished workload, which needs to be carried from one period to the next period, e.g., Cheung et al. (2002) and Linn et al. (2003).

The scheduling of stacking and retrieval operations of containers in the yard has to take the QC operations into account. More precisely, a horizontal transport vehicle which has to receive or to deliver a particular container at a yard block will arrive there at a certain point in time depending on the progress of QC operations. These arrival times of vehicles refer to ready times of stacking and retrieval operations in the YC scheduling. Further constraints arise if two YCs operate within a yard block.

Here, interference issues similar to the QCSP need to be considered for the YCs, see, e.g., Ng (2005) and Jung and Kim (2006).

3.3.3 Horizontal Transport

For the horizontal transport operations three decisions have to be made, which, in general, all aim at a high productive horizontal transport by minimizing the empty travel of vehicles:

- Assignment of vehicles to QCs
- Assignment of transport orders to vehicles and sequencing of assigned orders
- Routing of vehicles and traffic control

With the first decision, vehicles are either assigned exclusively to QCs or they are pooled where each vehicle serves different QCs. In case of an exclusive assignment, the potential for reducing empty travel of vehicles is largely limited because QCs typically operate in single cycle mode. Single cycle mode means that the crane either performs consecutive loading operations or consecutive unloading operations exclusively. Hence, an empty travel from the quay to the yard or from the yard to the quay is unavoidable, depending on whether containers are loaded or unloaded by the QC. Contrasting an exclusive assignment of vehicles to QCs, pooled vehicles serve more than one QC. If some QCs load containers while others unload containers, empty movement of vehicles can be reduced by performing loaded travels from the quay to the yard and vice versa. Avoiding empty movement of vehicles reduces the waiting time of QCs, see Böse et al. (2000). The increased crane productivity in turn accelerates the service of a vessel. Murty et al. (2005a,b) use the pooling strategy to minimize the number of vehicles required for the transport operations. They note that pooling comes along with an increased effort for dispatching vehicles to ensure that each crane receives sufficient transport capacity.

The second decision regarding the horizontal transport is the assignment of transport orders to vehicles. It is usually made on a short planning horizon of a few minutes to respect the actual progress of QCs in performing their operations. Since most vehicle types can carry only a single container at a time, the vehicles execute the assigned orders in a pickup and delivery fashion. The container sequences of the QCs and the progress of crane operations define ready times and due dates of pickup and delivery operations that have to be respected. Since horizontal transport operations are planned on a short-time basis, the sequencing of (the few) transport orders assigned to a vehicle is basically an easy to solve problem. However, difficulties arise from the need to provide solutions within a few seconds.

If AGVs are employed at a terminal a precise coordination between the vehicles and the QCs and YCs is required. Otherwise, so-called deadlocks can arise. According to Kim et al. (1997), a deadlock is a permanent blocking of concurrent processes for which the resource requests can never be satisfied. For example, a deadlock situation arises if an AGV waits at a yard block to get its loaded container unloaded but the dedicated YC already holds the next container to load onto this AGV. Deadlock conditions in automated CTs and methods to avoid or resolve them have been investigated among others by Kim et al. (2006) and Lehmann et al. (2006). The coordination of AGVs and crane operations becomes increasingly difficult if dual-load AGVs are considered. These AGVs can transport two 20′ containers at a time. To use this potential, pattern-based algorithms for assigning and sequencing the pickups and deliveries of containers are provided by Grunow et al. (2006). A simulation study reveals that more than 30% of the 20′ containers are transported using the dual-load capability. Extensive investigations concerning deadlock detection, deadlock prevention, and dual-load operations planning are provided by Lehmann (2006).

The third planning decision of horizontal transport operations is choosing the travel routes and controlling of traffic. These tasks are left up to the drivers in a manually operated CT. They usually take the shortest path to the next destination and are able to rearrange their vehicles at the apron to fit the container deliveries to the loading sequence of a QC. AGVs and ALVs are less flexible in route choice because they have to follow a preset traffic course. This leads to reduced potential for minimizing travel distances. The traffic control is usually part of the operating system of the vehicles and cannot be influenced directly by the CT operators.

3.4 Landside Operations Planning

Landside operations concern the service of trains and external trucks. Basically, these service operations are similar to the service operations of barges and container vessels at the quay. Serving trains requires either the usage of straddle carriers or the usage of gantry cranes spanning the rail tracks. Since rail tracks do not lead into the yard area, a horizontal transport of containers needs to be organized. Planning methods used for the horizontal transport at the seaside of the terminal can also be applied to the horizontal transport at the landside. The terminal operator is, however, more flexible in determining sequences for loading and unloading operations because containers are not stacked on trains. Consequently, a reduction potential exists for empty travel of vehicles and for reshuffling of containers in the yard. A further objective is the minimization of shunting activities of trains, see Steenken et al. (2004).

External trucks are served directly at the blocks, if YCs are used for the yard operations. Otherwise, they are served at a parking area, where a container transport from the yard to this area is required. Empty travel of terminal vehicles can be minimized if trucks delivering export containers and trucks picking up import containers are served in parallel. Customs clearance, documentation, and the travel of trucks within the terminal determines the arrival times at the yard blocks or at the parking area. Therefore, storage, retrieval, and transport processes are often planned not until trucks arrive at these locations. The limited potential for planning is a primary difference between landside operations and seaside operations. Since trucks receive even less priority than the other transport means, waiting times are wide-spread and more or less accepted. However, to provide a good service quality, the minimization of truck waiting time is an important objective of planning landside operations, see Kim et al. (2003). Further objectives concern the balancing of workload of YCs and the minimization of the number of cranes required for the service, see Froyland et al. (2008).

3.5 Workforce Planning

In a hardly automated CT the workforce planning is crucial in order to operate the terminal with skilled laborers. Two problems have to be solved here:

- Provision of workforce capacity
- Scheduling of labor tasks

With the first decision, it is decided on the workforce capacity that is needed to handle the workload of a terminal within a period. More precisely, a sufficient number of laborers to operate the equipment needs to be determined. A basic constraint is to respect the dynamic availability of the workforce, which follows from the implemented shift system. Legato and Monaco (2004) give an approach to the design of an appropriate shift system for a terminal. Agreements between labor unions and the terminal management restrict the potential decisions. In a second model they describe the short-term workforce utilization using additional laborers in situations of high workload, called peaks. Murty et al. (2005a,b) estimate the workload of the horizontal transport system on a shift basis and derive workforce requirements. Kim et al. (2004b) assign individual operators to QCs, YCs, and yard trucks with respect to given work plans and the laborer's skills.

In particular regarding the seaside operations, automated solutions are not yet applicable in practice. Moreover, related workforce aspects are seldom considered in the scientific literature. Operating a QC requires a group of laborers, called a *gang*. A gang consists of a skilled crane driver, stevedores for the lashing of containers, a foreman which coordinates the operations, and, possibly, a number of drivers for the horizontal transport means. Since QCs are usually manned for a complete work shift, a high utilization of manned QCs is a secondary objective of the terminal management besides service quality maximization. To account for this issue, Meisel and Bierwirth (2006) provide an approach aiming at the minimization of manned QC shifts within a combined berth allocation and QC assignment problem. However, with labor cost as the only objective, a reduction of the service quality is observed. A combined objective of both fields allows to balance between the profitability of the terminal and the satisfaction of terminal customers.

The second decision of workforce planning is the scheduling of labor tasks. Scheduling labor tasks is often within the scope of the above described equipment scheduling problems, such as QC and YC scheduling, and horizontal transport planning. Nevertheless, scheduling individual laborers is still necessary for handling of special containers, e.g., to connect reefer containers to electric supply in the yard, see Hartmann (2004). Objectives of labor task scheduling are, for example, the minimization of tardiness of task completions (Hartmann, 2004) or the minimization of the number of required laborers (Lim et al., 2004a).

Chapter 4
Related Work on Seaside Operations Planning

Due to the variety of technical equipments and terminal layouts, research has produced a multitude of optimization models for seaside operations planning in container terminals. This chapter provides literature surveys for the operations planning problems being in the focus of the thesis. Section 4.1 provides a classification scheme and a literature survey for BAP and QCAP formulations. Section 4.2 provides a classification scheme and a literature survey for QCSP formulations. Section 4.3 describes relationships of the seaside planning problems and well known Operations Research problems.

4.1 Related Work on the BAP and the QCAP

4.1.1 Classification Scheme

To show similarities and differences in the existing models for berth allocation, a classification scheme is developed in the following. Studies that concentrate on quay crane assignment either presuppose a particular type of BAP or integrate quay crane assignment decisions in the berth planning process. For this reason, QCAP approaches are captured by the classification scheme as well. Problems are classified according to four attributes. The *spatial attribute* concerns the berth layout and water depth restrictions. The *temporal attribute* describes the temporal constraints for the service process of vessels. The *handling time attribute* determines the way vessel handling times are considered in the problem. The fourth attribute defines a *performance measure* for evaluating possible solutions to a problem. Each attribute can take different values. They are listed in Table 4.1.

Spatial, temporal, and handling time attributes have been described in Sect. 3.2.1. The performance measures listed in Table 4.1 reflect different service quality criteria. Minimizing the waiting time or the handling time of a vessel aims at providing a competitive service to vessel operators. If both objectives are pursued (i.e., *wait* and

F. Meisel, *Seaside Operations Planning in Container Terminals*,
Contributions to Management Science,
DOI: 10.1007/978-3-7908-2191-8_4, © Physica-Verlag Berlin Heidelberg 2009

Table 4.1 A classification scheme for BAP formulations

Value	Description
1. Spatial attribute	
disc	The quay is partitioned in *disc*rete berths
cont	The quay is assumed to be a *cont*inuous line
hybr	The *hybr*id quay mixes up properties of discrete and continuous berths
draft	Vessels with a *draft* exceeding a minimum water depth cannot be berthed arbitrarily
2. Temporal attribute	
stat	In *stat*ic problems there are no restrictions on the berthing times
dyn	In *dyn*amic problems arrival times restrict the earliest berthing times
due	*Due* dates restrict the latest allowed departure times of vessels
3. Handling time attribute	
fix	The handling time of a vessel is considered *fix*ed
pos	The handling time of a vessel depends on its berthing *pos*ition
QCAP	The handling time of a vessel depends on the assignment of QCs
QCSP	The handling time of a vessel depends on a QC operation schedule
4. Performance measure	
wait	*Wait*ing time of a vessel
hand	*Hand*ling time of a vessel
compl	*Compl*etion time of a vessel
speed	*Speed*up of a vessel to reach the terminal before the expected arrival time
tard	*Tard*iness of a vessel against the given due date
order	Deviation between the arrival *order* of vessels and the service order
rej	*Rej*ection of a vessel
res	*Res*ource utilization effected by the service of a vessel
pos	Berthing of a vessel apart from its desired berthing *pos*ition
misc	*Misc*ellaneous

hand are set), the port stay time of vessels is minimized. Minimizing the completion times of vessels (*compl*) aims at earliest possible departures. In the presence of soft arrival times or soft due dates either the speedup of vessels (*speed*) or the tardiness of vessels (*tard*) has to be minimized. The *order* measure strives at a reduction of the deviation between the arrival order of vessels and the planned service order. It is assessed by the number of vessels not served in First-come First-served manner. If it is foreseeable that a vessel cannot be served within the desired time window, it may be rejected at the terminal (and possibly reassigned to another terminal of the port). Hence, the minimization of vessel rejections (*rej*) is considered as a goal in some models. If labor or other resources are scarce at a terminal, the resource utilization (*res*) is optimized. The minimization of deviations between chosen berthing positions and desired positions (*pos*) aims at reducing the travel distances for the horizontal transport vehicles. If none of the above performance measures is used in a BAP formulation, the value *misc* (miscellaneous) is set in the classification.

The above listed measures address criteria to be minimized. Either the minimization of the total measure for all vessels or the minimization of the measure for the worst performing vessel can serve as an objective function in a BAP. A total measure is denoted in the classification scheme by a $\sum()$ function and a worst performing measure, i.e., a min–max objective, is denoted by a max() function. Vessel specific weights are indicated by the denotation w. Moreover, if weights appear with an index, i.e., w_1 to w_4, they address weights of combined performance measures.

Using the introduced classification scheme, a certain type of BAP is described by a selection of values for each of the four attributes. As an example, consider a problem where the quay is partitioned into discrete berths serving the vessels exclusively (*disc*). The arrival times restrict the earliest berthing of vessels (*dyn*) and handling times are known and fixed (*fix*). The objective is to minimize the total cost arising for tardiness of vessels (*tard*) and for berthing vessels apart from desired berthing positions (*pos*). Different cost rates (w_1, w_2) apply to these performance measures. According to the proposed scheme, this problem is classified by $disc \mid dyn \mid fix \mid \sum(w_1\ tard + w_2\ pos)$. In case that the maximum tardiness of vessels has to be minimized, the problem is classified by $disc \mid dyn \mid fix \mid max(tard)$.

4.1.2 Problem Classification

Table 4.2 gives a comprehensive survey of berth allocation and quay crane assignment formulations from the literature. Some authors outline approaches more or less informally while others provide precise optimization models. If a unique classification of a paper is not possible according to the given information, the best fit of classifying attributes is taken. The classification exclusively covers research dealing with the operational decisions regarding the BAP and QCAP. Not covered are studies employing analytical models, simulation, and queuing theory as is used for the evaluation of investment decisions and berthing policies, and for the determination of terminal throughput and system dynamics of CTs (see Edmond and Maggs, 1978; Schonfeld and Frank, 1984; Lai and Shih, 1992; Legato and Mazza, 2001; Henesey et al., 2004; Dragovic et al., 2005, 2006). Papers containing ideas and results published elsewhere, for example Kim (2005) and Crainic and Kim (2007), are also excluded.

As shown in Table 4.2, discrete and continuous problems are almost in balance, while dynamic problem formulations clearly prevail against static ones. The handling times of vessels are assumed to depend on the berthing positions in almost every discrete BAP formulation, because they are easily assessable in discrete models. However, only two continuous formulations and one half of the hybrid BAP formulations care for position based handling times. The QC resource is considered only in a few BAP formulations. Most models aim at the minimization of the port stay time of vessels. Frequently addressed are also the minimization of tardy vessel departures and berthing positions different from desired berthing positions.

Table 4.2 Overview of BAP formulations

Problem classification	Reference
$disc \mid stat \mid pos \mid \sum(wait+hand)$	Imai et al. (2001)
	Hansen and Oğuz (2003)
$disc \mid stat \mid pos \mid \sum(wait+hand+w_1\ order)$	Imai et al. (1997)
$disc \mid stat, due \mid pos \mid \sum w\ rej$	Imai et al. (2008b)
$disc \mid stat \mid pos, QCSP \mid \sum(wait+hand)$	Lee et al. (2006)
$disc \mid dyn \mid pos \mid \sum(wait+hand)$	Imai et al. (2001)
	Hansen and Oğuz (2003)
	Monaco and Sammarra (2007)
$disc \mid dyn \mid pos \mid \sum w(wait+hand)$	Imai et al. (2003)
	Theofanis et al. (2007a)
$disc \mid dyn \mid pos \mid \sum(w_1\ wait+w_2\ tard+w_3\ pos)$	Hansen et al. (2008)
$disc \mid dyn \mid pos \mid \sum w\ tard$	Golias et al. (2006)
$disc \mid dyn \mid pos \mid \sum w\ tard, \sum(wait+hand)$	Imai et al. (2007b)
$disc \mid dyn \mid pos \mid misc$	Golias et al. (2007)
$disc \mid dyn, due \mid pos \mid \sum w(wait+hand)$	Cordeau et al. (2005)
	Mauri et al. (2008)
$disc \mid dyn, due \mid pos \mid \sum w\ rej$	Imai et al. (2008b)
$disc, draft \mid dyn \mid pos \mid \sum(wait+hand)$	Han et al. (2006)
$disc, draft \mid dyn, due \mid pos \mid \sum wait$	Zhou et al. (2006)
$disc \mid dyn \mid pos, QCAP \mid \sum(wait+hand)$	Imai et al. (2008a)
$disc \mid dyn \mid QCAP \mid \sum(wait+hand+tard)$	Liang et al. (2009)
$disc \mid dyn, due \mid QCAP \mid -\sum(w_1 res - w_2 pos)$	Giallombardo et al. (2008)
$cont \mid stat \mid fix \mid max(compl)$	Li et al. (1998)
$cont \mid stat \mid fix \mid \sum w\ compl$	Guan et al. (2002)
$cont \mid stat \mid QCAP \mid \sum(w_1\ wait+w_2\ speed + w_3\ tard + w_4\ pos)$	Park and Kim (2003)
	Rashidi (2006)
$cont \mid stat \mid QCAP \mid max(compl)$	Oğuz et al. (2004)
$cont \mid dyn \mid fix \mid \sum w(wait+hand)$	Guan and Cheung (2004)
$cont \mid dyn \mid fix \mid \sum(w_1\ wait+w_2\ pos+w_3\ rej)$	Wang and Lim (2007)
$cont \mid dyn \mid fix \mid \sum(w_1\ tard+w_2\ pos)$	Moon (2000), Park and Kim (2002)
	Kim and Moon (2003)
	Briano et al. (2005)
$cont \mid dyn \mid fix \mid max(res)$	Lim (1998)
$cont, draft \mid dyn \mid fix \mid max(res)$	Lim (1999), Tong et al. (1999)
	Goh and Lim (2000)
$cont \mid dyn \mid pos \mid \sum(wait+hand)$	Imai et al. (2005)
$cont \mid dyn \mid QCAP \mid \sum res$	Meisel and Bierwirth (2006)
$cont \mid dyn \mid QCAP \mid \sum w\ res$	Theofanis et al. (2007b)
$cont \mid dyn \mid QCAP, QCSP \mid max(tard)$	Liu et al. (2006)
$cont \mid dyn \mid QCAP, QCSP \mid max(compl), \sum(wait+hand)$	Meier and Schumann (2007)
$cont \mid dyn \mid QCAP, QCSP \mid \sum(wait+hand+w_1\ tard)$	Ak and Erera (2006)
$cont \mid dyn, due \mid QCAP \mid max(res)$	Hendriks et al. (2008)
$cont \mid dyn, due \mid QCAP \mid \sum hand+w_1\ res$	Legato et al. (2008)
$hybr \mid dyn \mid fix \mid \sum(w_1\ wait+w_2\ pos)$	Moorthy and Teo (2006)
	Dai et al. (2008)
$hybr \mid dyn, due \mid fix \mid \sum w\ pos$	Chen and Hsieh (1999)
$hybr \mid dyn \mid pos \mid \sum(wait+hand)$	Imai et al. (2007a)
$hybr \mid dyn, due \mid pos \mid \sum w(wait+hand)$	Cordeau et al. (2005)
$hybr, draft \mid dyn \mid pos \mid \sum(wait+hand)$	Nishimura et al. (2001)
$hybr, draft \mid dyn \mid pos \mid max(compl), \sum wait, \sum order$	Cheong et al. (2007)
$hybr, draft \mid dyn \mid pos \mid \sum pos$	Hoffarth and Voß (1994)
$hybr \mid dyn \mid pos, QCAP \mid \sum wait, \sum tard$	Lokuge and Alahakoon (2007)

In the following, the relevant papers are briefly reviewed according the grouping into discrete, continuous, and hybrid problems. Formulations with handling time characteristics *fix* or *pos* are merely summarized, whereas formulations that consider the crane resource within berth planning are presented in more detail.

4.1.2.1 Discrete Problems

The discrete BAP has been studied in the static and in the dynamic variant by Imai et al. (2001). In both problem variants the assignment and sequencing of vessels to berths is searched with respect to minimum waiting and handling times of the vessels. A Lagrangean relaxation based heuristic is presented to solve the problem. Hansen and Oğuz (2003) and Monaco and Sammarra (2007) provide more compact MIP formulations for the same problem. In the discrete static BAP considered by Imai et al. (1997) not only waiting and handling times of the vessels are minimized but also the deviation between the arrival order of vessels and the service order. The problem is reduced to a classical assignment problem, which is solved by the Hungarian method. In a recent paper of Imai et al. (2008b) the minimization of the weighted number of vessel rejections is handed over to a Genetic Algorithm (GA). A vessel is rejected at a terminal if it cannot be served without overshooting a due date, represented by a maximum acceptable waiting time.

Dynamic variants of the discrete BAP are considered by Imai et al. (2003) and Theofanis et al. (2007a) in the context of weighted port stay times for the vessels. In the problem considered by Hansen et al. (2008) not only handling times but also service costs of vessels depend on the berth they are assigned to. Pursued is a tardiness objective which accounts for departure time related costs including penalties for tardiness as well as benefits for early departures. A Variable Neighborhood Search, used to solve the problem, turns out to be superior to the GA of Nishimura et al. (2001). Further departure time related objectives for the discrete dynamic BAP are proposed by Golias et al. (2006) and Imai et al. (2007b). In the model of Golias et al. (2007) arrival times and handling times of vessels are considered as stochastic variables. Since no specific performance measure is supposed for the problem, it is classified miscellaneous. A discrete dynamic BAP with due dates is formulated by Cordeau et al. (2005). A Tabu Search method is presented that outperforms the First-come First-served rule and also CPLEX. Mauri et al. (2008) design a Column Generation approach for the problem of Cordeau et al. (2005) which delivers better solutions in shorter runtime than Tabu Search. In the models of Han et al. (2006) and Zhou et al. (2006) the draft of vessels restricts the berth assignment decisions. In both papers, a GA is proposed to solve the problem. Moreover, Zhou et al. (2006) consider stochastic arrival and handling times of vessels and a waiting time restriction that is classified as a due date.

The crane resource is considered within a discrete BAP by Lee et al. (2006), Imai et al. (2008a), Liang et al. (2009), and Giallombardo et al. (2008). Lee et al. (2006) develop a GA to obtain berth plans that are evaluated by generating a feasible work plan for a given number of cranes at each berth. The evaluation bases on a modification of the QCSP model of Kim and Park (2004), which is not solved in the paper. Instead, a small-sized instance is provided as example. In the paper of Imai et al. (2008a) it is supposed that a certain number of QCs has to be assigned to each vessel. The model of Liang et al. (2009) decides on the assignment of cranes to berths as well as on the berthing times and the positions of the vessels. In Giallombardo et al. (2008), cranes, berthing times, and berthing positions are assigned to vessels, such that the crane utilization (*res*) is maximized and the berthing position dependent container flow between pairs of vessels (*pos*) is minimized.

4.1.2.2 Continuous Problems

The continuous static BAP with fixed vessel handling times has been introduced by Li et al. (1998). The problem is formulated as a "multiple-job-on-one-processor" scheduling problem. This allows to adapt the First-Fit-Decreasing heuristic, well-known from Bin Packing, for minimizing the maximum completion time among the vessels. Guan et al. (2002) propose to minimize the total weighted completion time of vessels for this type of problem and provide a priority rule based heuristic.

The continuous dynamic BAP with fixed handling times has been investigated in a number of studies. Guan and Cheung (2004) develop a tree search procedure to minimize the total weighted port stay time of vessels. In the problem of Wang and Lim (2007) the minimization of penalty cost for rejected vessels and apart berthing positions are the pursued objectives. A stochastic beam search algorithm is presented that is capable to solve instances with up to 400 vessels. A further objective, namely the minimization of tardiness of vessels, is treated by Moon (2000), Park and Kim (2002), Kim and Moon (2003), and Briano et al. (2005). Several solution methods are proposed for this problem, including a sub-gradient method (Park and Kim, 2002) and a Simulated Annealing approach (Kim and Moon, 2003). Lim (1998) formulates a problem, where the berthing times of vessels are already set by the arrival times. Instead, suitable berthing positions need to be determined and the goal is to minimize the maximum quay length required to serve vessels in accordance with the schedule. In this formulation of the problem, the goal is classified as a resource objective. The approach of Lim (1998) has been continued by Lim (1999), Tong et al. (1999), and Goh and Lim (2000).

The continuous BAP with handling times depending on berthing positions is studied by Imai et al. (2005). The authors suggest a heuristic solution method, which solves a discrete BAP first and then improves the obtained solution by shifting vessels along the quay as allowed in the continuous BAP.

The continuous BAP in combination with crane assignment issues has been introduced by Park and Kim (2003). In this pioneering model it is simultaneously decided on berthing times, berthing positions, and the assignment of QCs to vessels at first. Here, the idea is to assign single QC-hours to the vessels such that variable-in-time QC-to-Vessel assignments are possible. Next, it is decided on the specific cranes that serve a vessel. A Lagrangean relaxation based heuristic is used at the first decision level and dynamic programming is applied at the second level. A combined model for both decision levels is presented by Rashidi (2006).

The continuous static BAP with crane assignment issues is also studied by Oğuz et al. (2004). The pursued objective is to minimize the maximum completion time among the vessels. In contrast to Park and Kim (2003), Oğuz et al. (2004) merely consider time-invariant QC-to-Vessel assignments. However, crane productivity losses, which are frequently observed in practice, are anticipated in the made assignments. Unfortunately, the power of both approaches is not yet compared in the literature.

An approach aiming at the improvement of crane utilization is provided by Meisel and Bierwirth (2006). Here, a set of promising QC-to-Vessel assignment patterns is generated for each vessel at first and then a priority rule is used to fix the berthing time, the berthing position, and the particular crane assignment pattern for every vessel. The minimization of cost for manning QCs is pursued. On this basis, several real-world berth plans are considerably improved without deteriorating the service quality substantially. Also Theofanis et al. (2007b) aim at an effective use of cranes by penalizing crane assignments that do not meet a targeted productivity rate. As a common solution method, a GA is proposed for the problem, which does not deliver satisfying results yet.

In a couple of papers, crane assignment and crane scheduling are involved to estimate vessel handling times for the berth planning. Liu et al. (2006) propose to derive vessel handling times from QCSP schedules which are generated in a preprocessing. Berthing times and QC-to-Vessel assignments are subsequently determined for the vessels with respect to given berthing positions and the projected handling times. The minimization of the maximum relative tardiness of vessel departures is considered as objective. Meier and Schumann (2007) generate a berth plan using the approach of Guan and Cheung (2004). Next, QC schedules are build for all vessels basing on the QCSP model of Zhu and Lim (2006). The gained handling times are used to revise the prior berth plan iteratively, which is controlled by a Multi-Agent System (MAS). However, the conducted computations indicate that this coordination is not yet effective in minimizing service objectives. A monolithic model for continuous dynamic berth allocation, crane assignment, and crane scheduling is presented by Ak and Erera (2006). The captured subproblems are considered on a highly aggregated level, which enables applying Tabu Search for the minimization of the port stay time and the avoidance of penalty cost for tardy departures of the vessels.

Further approaches include due dates of vessels into a dynamic BAP, which restricts the service of vessels to time windows. Hendriks et al. (2008) consider berth allocation at a tactical level, where service time windows, quay space, and crane capacity are reserved for a set of periodically arriving vessels. The objective is to minimize the maximum crane capacity reserved for a period. Legato et al. (2008) propose to take berthing positions, berthing times, and due dates from a berth plan that is generated using the model of Park and Kim (2003). The task remaining is to assign the available cranes to vessels such that the total handling time of vessels and the number of utilized cranes are minimized. A heuristic procedure is sketched for this problem and demonstrated at an example.

4.1.2.3 Hybrid Problems

The hybrid BAP with fixed handling times is investigated by Moorthy and Teo (2006), Dai et al. (2008), and Chen and Hsieh (1999). In Moorthy and Teo (2006) berthing areas of vessels are determined at a tactical level. The goal is to achieve robust berth plans with respect to stochastic perturbations of vessel arrivals. To identify the impact of vessel delays, the service processes of vessels are considered as activities and represented in a precedence graph which is analyzed using the Project Evaluation Review Technique. Dai et al. (2008) transfer the gained results to the operational level, where precise positions are searched within the projected berthing areas. For this problem a Simulated Annealing algorithm is proposed which minimizes vessel waiting time and the berthing position dependent container flow between pairs of vessels. In an early paper of Chen and Hsieh (1999), a MIP formulation is given for a same problem which also incorporates vessel due dates.

Hybrid BAP formulations with position dependent vessel handling times are studied from different perspectives in several papers. Imai et al. (2007a) investigate a hybrid BAP for indented berths whereas Cordeau et al. (2005) derive a hybrid problem from a discrete BAP where the quay is dynamically repartitioned. Additionally, the models of Nishimura et al. (2001), Cheong et al. (2007), and Hoffarth and Voß (1994) incorporate the draft of vessels. Although crane assignment issues are approached in Hoffarth and Voß (1994), a heuristic is proposed that does not involve crane assignments in the berth allocation planning. Crane assignment decisions are explicitly included in a hybrid BAP by Lokuge and Alahakoon (2007) and accompanying papers, see Lokuge and Alahakoon (2004, 2005) and Lokuge et al. (2004). In this problem, multiple vessels can be served at one berth at a time. A set of cranes, shared by simultaneously served vessels, is assigned to each berth. A MAS is designed where all decisions to be made are distributed among specialized software agents. The system performance is compared with an existing terminal operating system (Jaya Container Terminal, Sri Lanka), achieving considerable reductions regarding vessel waiting times and tardiness.

Concluding the classification of BAP formulations, it can be seen that incorporating decision dependent handling times of vessels is one of the most active streams of current research in berth allocation planning. The impact of berthing positions on handling times is well-established in discrete and hybrid BAP formulations, but it is hardly considered in the continuous case. This is surprising because a deviation between a vessel's berthing position and the storage position of export containers in the yard certainly impacts the handling time in the continuous case as well. The consideration of the QC resource for the determination of handling times is a driving force of current research. Papers published on this issue reveal the growing interest in the interdependencies between seaside planning problems.

4.2 Related Work on the QCSP

4.2.1 Classification Scheme

As for berth planning problems, there is no classification scheme existing for QC scheduling problems so far. The proposed scheme classifies problems according to four attributes. The *task attribute* concerns the definition of tasks that represent the workload of the considered vessel. The *crane attribute* describes the availability of QCs at the vessel and the consideration of the crane movement speed. The *interference attribute* addresses the spatial constraints that are defined in a problem. These attributes have been described in detail in Sect. 3.2.3. A fourth attribute defines the *performance measure* for evaluating solutions to a problem. Each of the four attributes can take different values. They are listed in Table 4.3.

Most of the performance measures for crane schedules aim at short vessel handling times to allow for earliest possible departures. Minimizing the completion time of tasks (*compl*) serves the purpose of a short vessel handling time. The makespan of a schedule, i.e., the maximum completion time among all tasks, is of particular interest, because it determines the departure time for a vessel. An effective utilization of cranes leads also to short vessel handling times. Related performance measures are

Table 4.3 A classification scheme for QCSP formulations

Value	Description
1. Task attribute	
Area	Tasks refer to *area*s of bays
Bay	Tasks refer to single *bays*
Stack	Tasks refer to container *stack*s within a bay
Group	Tasks refer to *groups* of containers
Container	Tasks refer to single *container*s
prmp	Preemption of tasks is allowed
prec	Precedence relations among tasks are given
2. Crane attribute	
ready	Individual *ready* times of QCs are given
TW	Cranes are available at the vessel within hard time windows
pos	Initial (and final) *pos*itions of QCs are prescribed
move	Travel time for crane *move*ment is respected
3. Interference attribute	
cross	Non-*cross*ing of QCs is respected
save	*Saf*ety margins between QCs are respected
4. Performance measure	
compl	*Compl*etion time of a task
finish	*Finish*ing time of a QC
util	*Util*ization rate of a QC
through	*Through*put of a QC
move	*Move*ment of a QC

the finishing time of cranes (*finish*), the utilization rate of cranes (*util*), the throughput of cranes (*through*), and the time spent for moving cranes to other quay positions (*move*).

Like in the classification of berth allocation problems, a $\sum()$ function indicates a total performance measurement of either tasks or QCs, and a $\max()$ function indicates a worst performance measurement. Objective functions with weighted tasks or cranes are indicated by a weight w, whereas weights w_1, w_2, etc., address a multi-objective function. As an example for the classification scheme consider a problem where the tasks consist of all loading and unloading operations of containers in bays. Assume further that initial positions and a moving speed are given for the cranes and non-crossing as well as safety margins must be respected. If the objective is to minimize the completion of the latest finished task, the problem is referred to as $Bay \mid pos, move \mid cross, save \mid max(compl)$ in the classification.

4.2.2 Problem Classification

Table 4.4 gives a comprehensive survey of QCSP formulations from the literature. The classification comprises optimization models and verbally introduced problem descriptions by a best fit. The survey shows that most approaches define tasks on the basis of single bays or container groups. Non-crossing constraints are involved in the majority of research, whereas safety margins have much less been considered. Crane attributes are neglected in diverse studies, i.e., there are no ready times or time windows and also no movement speed given for the cranes. Most of the published QCSP formulations are dealing with the makespan criterion. The following survey takes up the distinction of models by the task attribute. Crane scheduling problems for complete bays and for container groups are considered next. The remaining approaches are presented together in a third subsection.

4.2.2.1 Scheduling of Complete Bays

In the QCSP with complete bays it is searched for a crane schedule where each bay is exclusively served (usually without preemption) by one QC. For a realistic consideration of this problem, the incorporation of the non-crossing condition is more or less inevitable. Basically, non-crossing of cranes is quite easy to assure by generating schedules, where all QCs have an identical moving direction along the vessel which is not changed during the service. Such schedules are referred to as unidirectional schedules in the following.

Under the premise of unidirectional crane schedules, some results on approximation algorithms are presented for the QCSP in Lim et al. (2004c). Moreover, a reformulation of the considered QCSP as a constraint programming model is provided in Lim et al. (2004d). The authors show that every unidirectional schedule can be obtained from a certain assignment of tasks to QCs. Hence, the problem

Table 4.4 Overview of QCSP formulations

Problem classification	Reference
$Area \mid - \mid cross \mid max(util)$	Winter (1999)
	Steenken et al. (2001)
$Area \mid - \mid cross, save \mid -\sum w\, through$	Lim et al. (2002, 2004b)
$Bay \mid - \mid cross \mid max(compl)$	Lim et al. (2004c,d, 2007)
	Zhu and Lim (2006)
	Lee et al. (2007, 2008a)
$Bay \mid - \mid cross \mid \sum w\, compl$	Lee et al. (2008b)
$Bay \mid move \mid cross \mid \sum move$	Ak and Erera (2006)
$Bay \mid pos, move \mid cross, save \mid max(compl)$	Liu et al. (2006)
$Bay, prmp \mid - \mid - \mid \sum w\, compl$	Daganzo (1989)
	Peterkofsky and Daganzo (1990)
$Bay, prmp \mid pos, move \mid cross, save \mid max(compl)$	Liu et al. (2006)
$Stack, prec \mid - \mid - \mid max(compl)$	Goodchild and Daganzo (2004)
	Zhang and Kim (2009)
$Group, prec \mid ready, pos, move \mid cross, save \mid w_1\, max(compl) + w_2 \sum finish$	Kim and Park (2004)
	Moccia et al. (2006)
	Sammarra et al. (2007)
$Group, prec \mid ready, pos, move \mid cross, save \mid \sum w\, finish + \sum w\, tard$	Tavakkoli-Moghaddam et al. (2009)
$Group, prec \mid move \mid cross \mid max(compl)$	Ng and Mak (2006)
$Group, prec \mid TW \mid cross, save \mid max(compl)$	Jung et al. (2006)
$Container, prec \mid - \mid - \mid max(compl)$	Meisel and Wichmann (2008)

can be solved by exploring the space of task-to-QC assignments for which a constraint propagation method and a Simulated Annealing algorithm are presented. Recently, Lim et al. (2007) have shown for the QCSP with complete bays that there is always an optimal schedule among the unidirectional ones. This seminal result demonstrates that searching the space of unidirectional schedules is not a heuristic reduction of the problem. An exact method for the unidirectional problem delivers the optimal solution even if the premise of unidirectional crane schedules is dropped.

Liu et al. (2006) propose a MIP model for the QCSP which includes initial crane positions, moving speed, and interference conditions for the cranes. The structure of unidirectional schedules is anchored in this model, which allows to formulate the non-crossing condition and safety margins in a straight-forward manner. Furthermore, with the focus put on unidirectional schedules, the search space is significantly reduced which allows to solve non-trivial instances by a standard solver.

In a couple of papers, schedule unidirectionality is neither assumed in models nor in algorithms. A Branch-and-Bound method with limited capability and a better performing Simulated Annealing algorithm are presented for the QCSP with complete bays by Zhu and Lim (2006). A GA and a greedy algorithm are developed by Lee et al. (2007, 2008a) for the same problem. The approach is augmented in Lee et al. (2008b) by replacing the makespan criterion with the total weighted completion time of tasks. Ak and Erera (2006) treat the QCSP for a set of vessels that are served simultaneously at the quay. The assignment problem of cranes to the bays of the vessels is modeled as a min-cost flow problem, with the QC travel time between vessels to be minimized.

Scheduling of QCs with preemption allowed has been investigated by Daganzo (1989). The idea is to assign cranes to bays for certain time slots, such that the overall workload is well balanced for the cranes. As a consequence, a bay might be served consecutively by different QCs. Note that this early work does not take crane interference into consideration. The goal is to minimize the total weighted completion times of vessels. The considered scheduling problem is solved by rules of thumb and by a Branch-and-Bound method later proposed by Peterkofsky and Daganzo (1990). Preemptive schedules are also allowed in another version of the above mentioned model of Liu et al. (2006). The authors use this model to prove by experiment that the sharing of bays among cranes can significantly improve the makespan of a QCSP instance.

4.2.2.2 Scheduling of Container Groups

Enabling the cranes to share the workload of bays is the most typical feature of the QCSP with container groups. For this reason the need to avoid crane crossing is even more stressed in this problem than in the QCSP with complete bays. The QCSP with tasks defined on the basis of container groups has been introduced by Kim and Park (2004). Their model considers QC operations in detail by taking precedence relations among tasks, crane attributes, and crane interference into account. In this model, safety margins between QCs are enforced by a non-simultaneity constraint between tasks located in adjacent bays. The pursued objective is the minimization of the weighted sum of makespan and QC finishing times. The authors propose a Branch-and-Bound method and a Greedy Randomized Adaptive Search Procedure (GRASP). The Branch-and-Bound method outperforms GRASP in terms of solution quality but fails for larger test problems.

The model of Kim and Park (2004) has been refined by Moccia et al. (2006), leading to a more stringent problem formulation. They develop a Branch-and-Cut algorithm which significantly improves solutions for the benchmark suite provided in Kim and Park (2004). Sammarra et al. (2007) present a Tabu Search algorithm for the same problem where a neighborhood is defined by resequencing the tasks of a crane and by swapping tasks between cranes. Compared to the Branch-and-Cut of Moccia et al. (2006), Tabu Search cuts down the computation time significantly for the larger instances of the benchmark suite at the expense of a slightly weaker solution quality. Tavakkoli-Moghaddam et al. (2009) extend the model of Kim and Park (2004) towards scheduling cranes for a set of vessels in parallel. A GA is presented for the minimization of the total weighted finishing time of QCs and the total weighted tardiness of vessels (as known from berth planning). Unfortunately, the GA's performance is not compared with other algorithms proposed in this stream.

Ng and Mak (2006) obtain a QCSP with container groups by including the import containers and the export containers of a bay into two separate groups. A precedence relation is inserted between the two tasks of each bay. The problem is solved by partitioning the set of bays into areas such that sharing of workload by cranes is only possible at the border of two areas. Jung et al. (2006) discuss using time windows for the service of cranes but do not outline this feature in detail.

4.2.2.3 Further Problems

Taking a look at Table 4.4, it can be seen that defining tasks on the basis of bay areas, container stacks, or single containers is rather seldom dealt with in the scientific literature. Winter (1999) and Steenken et al. (2001) assign bay areas to QCs such that the maximum difference regarding the utilization of any two cranes is minimized. The authors show that crane scheduling on the basis of bay areas leads to a partitioning problem that can easily be solved optimal for instances of practical size. Lim et al. (2002) and Lim et al. (2004b) assign bay areas to QCs assuming individual throughput rates for the cranes. They aim at the maximization of the total throughput and propose several heuristics, where a Squeaky Wheel Optimization method combined with a local search performs best. Note that none of the mentioned approaches takes detailed crane schedules into consideration.

A stack-based QCSP model has been introduced by Goodchild and Daganzo (2004) and is further studied in a number of papers by Goodchild and Daganzo (2005a,b, 2006, 2007) and Goodchild (2006). The basic idea is to consider one crane processing the container stacks of one bay. Two precedence-related tasks are defined for each container stack, one for unloading the stack and one for loading the stack. Note that the loading and unloading of different stacks can be parallelized in the crane schedule, which is referred to as double cycling. The problem is to find a sequence for processing the stacks, which minimizes the makespan. The authors reformulate this problem as a two-machine flow shop scheduling problem which is solved to optimality using the rule of Johnson (1954). The approach is continued by Zhang and Kim (2009) who modify Johnson's rule in order to handle hatches that cover adjacent stacks. Double cycling is also addressed by Meisel and Wichmann (2008), who deal with crane scheduling on the basis of single containers. Contrasting the approach of Goodchild and Daganzo, reshuffle containers can be repositioned in the bay instead of temporarily unloading them, which accelerates the service process. A GRASP heuristic is used to solve the resulting scheduling problem.

QC scheduling is also considered within other operations planning problems concentrating, e.g., on stowage planning, horizontal transport operations, and yard crane scheduling, cf. Gambardella et al. (2001), Bish (2003), Kim et al. (2004a), Lee et al. (2005), Imai et al. (2006), Chen et al. (2007), and Canonaco et al. (2008). Since the primary focus of this research is not on the determination of crane schedules, these approaches are not explicitly addressed in the classification.

Concluding this review, major streams of QC scheduling research define tasks by complete bays or by container groups. Crane scheduling on the basis of container stacks or single containers, as investigated in the studies of Goodchild and Daganzo, Zhang and Kim (2009), and Meisel and Wichmann (2008), considers merely a single QC. This problem reduction eliminates important issues such as the assignment of tasks to cranes and the consideration of crane interference. Models that incorporate crane interference issues mainly address the non-crossing requirement. Safety margins and crane attributes like ready times, crane movement, etc., are often ignored. While one of these characteristics may influence the handling time of a vessel only moderately, their joint impact on vessel handling times is expected to be significant. Therefore, formulating rich QCSP models is crucial for deriving reliable QC schedules and vessel handling times.

4.3 Related OR Problems

Several well known problems from the field of Operations Research (OR) are closely related to the BAP, the QCAP, and the QCSP. In this section relations to machine scheduling, two-dimensional packing, project scheduling, and vehicle routing are briefly described. Reformulating the terminal specific problems to one of the standard problems enables existing solution methods to be applied. The potentials and limitations of such reformulations are briefly discussed.

In *machine scheduling* a set of jobs is considered which has to be processed by a set of machines. Job operations have to be sequenced on machines and feasible starting times for the job operations need to be found. A wide range of machine scheduling problems has been formulated in the literature, differing, for example, in the number and type of machines and in the particular machine routings required for the jobs. Usually a machine can process only one job at a time and job processing is assumed to be non-preemptive. Additionally, release times and due dates of jobs can be stated. Among others, makespan minimization is a typical objective. For an overview on this field of research the reader is referred to Pinedo (2002).

Machine scheduling approaches are of interest because berth allocation problems can be viewed as such, where the quay takes the role of a processor and the vessels take the role of the jobs. In the study of Li et al. (1998) a single processor (the whole quay) is able to handle several jobs in parallel. In the study of Guan et al. (2002) the quay is represented by a set of processors (quay segments) where a job is handled by a subset of processors simultaneously. Both studies address the continuous static BAP with fixed handling times. Release times of jobs have to be introduced to obtain a dynamic variant of this problem. The discrete BAP is treated as a scheduling problem for unrelated parallel machines with additional constraints by Imai et al. (1997) and Monaco and Sammarra (2007). In these approaches each berth is represented by one machine that processes the assigned jobs sequentially. Oğuz et al. (2004) view the QCs as a set of processors to incorporate the QCAP within the BAP. This allows for application of a parallel processor scheduling algorithm which strives for a makespan minimization.

Goodchild and Daganzo (2005b) formulate a stack-based QCSP as a two-machine flow shop problem. Each job (representing a container stack within a bay) consists of an unloading and a loading operation. One machine processes all unloading operations and the other machine processes all loading operations. Johnson's rule is applicable because the jobs have to pass the machines in identical order. Peterkofsky and Daganzo (1990) and Lim et al. (2007) discuss relations between bay-based QCSPs and the scheduling problem with parallel machines. While machines are basically unrelated in parallel machine scheduling, Lim et al. (2007) argue that quay cranes are related because the non-crossing requirement must be respected.

Two-dimensional packing deals with the arrangement of a set of rectangles with fixed size within an open-ended rectangular bin, see Baker et al. (1980). In a feasible solution no rectangles overlap. The objective is the minimization of the height of the packing within the bin.

The continuous static BAP with fixed vessel handling times shows close relations to two-dimensional packing problems. This becomes obvious by the space–time representation of berth plans. For berth planning only orthogonal packings need to be considered, i.e., each rectangle is positioned such that the edge representing the length of the vessel is in parallel to the axis representing the quay. Lim (1998) formulates the continuous BAP as a packing problem. In this model the berthing times of vessels are set according to the arrival times, and it is assumed that the quay is of infinite length. Berthing positions have to be determined such that the maximum quay length occupied at any time, i.e., the height of the packing, is minimized. This objective, however, is of little practical relevance because the length of a CT's quay is constant and berthing of vessels must be postponed whenever the quay space does not allow for a simultaneous service. The continuous dynamic BAP is modeled as a rectangle packing problem by Dai et al. (2008). In contrast to standard formulations of packing problems, the authors have to introduce additional constraints to respect arrival times of vessels.

Among the diverse project scheduling formulations, the *Resource Constrained Project Scheduling Problem* (RCPSP) is of predominant interest. Here, a set of activities has to be scheduled, each of them requiring a certain amount of one or more resources for their execution, see, e.g., Kolisch (1995). The available quantity of a resource is divided among activities executed simultaneously. Precedence constraints are typically involved in the RCPSP to express the relative order in which activities must be executed. The objective is the minimization of the project makespan.

A project scheduling approach is proposed by Moorthy and Teo (2006) to determine suitable berthing positions of vessels on a tactical level. They first generate berth plans and then state precedence relations between vessels occupying the same quay space while being served one after the other. Afterwards, the Project Evaluation Review Technique is used to determine the expected delays of vessels in a stochastic environment. In Meisel and Bierwirth (2006) the QC resource is considered within the BAP. Vessels, which are considered as activities in this approach, can be served in different modes, each one representing a different resource consumption pattern (QC-to-Vessel assignment). A precedence related dummy activity is introduced for each vessel activity in order to ensure that no vessel is served before its arrival time. A priority rule based method is used to schedule the activities while minimizing the QC resource utilization of the berth plan.

The *Vehicle Routing Problem* (VRP) deals with the assignment of delivery customer orders to a set of vehicles and with the determination of a route for each vehicle, starting at a depot, visiting the assigned customers, and returning to the depot. The capacity constraint of each vehicle is not allowed to be violated by the transport volume of its assigned orders. Contrasting other standard problems described above, which mostly pursue a min–max objective like makespan minimization, the VRP aims at minimizing a total measure, namely the overall route length of vehicles.

As demonstrated by Cordeau et al. (2005), the discrete BAP can be formulated as a multi-depot VRP with time windows. In this model each vessel is represented by a customer order and each berth is represented by a depot. A single vehicle is dedicated to each depot. A time window for the delivery of an order results from the expected arrival and departure time of the corresponding vessel. The objective of total route length minimization is replaced by the CT objective of vessel waiting time and handling time minimization.

Like the BAP, also the QCSP can be formulated as a variant of the VRP. Moccia et al. (2006) apply solution techniques for the Precedence Constrained Traveling Salesman Problem to crane scheduling. Sammarra et al. (2007) treat the QCSP as a VRP where adaptations are required to respect precedence constraints between QC tasks as well as crane interference issues. In both studies the VRP objective is replaced by a QCSP objective, namely the minimization of the makespan.

Summarizing, standard OR problems often nest inside seaside operations planning problems, but a one-to-one reformulation is not possible in most of the cases. Operations planning in CTs typically involves additional constraints and pursues different objectives. E.g., the minimization of the makespan, as typically pursued in machine scheduling, is not appropriate for berth planning, because it does not differentiate the individual service quality provided to vessels. Suitable formulations of the BAP, the QCAP, and the QCSP may be inspired by formulations of standard problems, but adaptations are still required. For this reason, standard solution methods cannot be applied in a straightforward manner.

Chapter 5
Integration Concepts for Seaside Operations Planning

In this chapter different concepts for an integrated solution of seaside planning problems are discussed. Section 5.1 assesses a sequential solution process of the focused problems and provides a theoretical framework for an integrated solution of optimization problems. Following these ideas, a survey of published integration concepts for the seaside planning problems is provided in Sect. 5.2. The particular integration concept to investigate in the thesis is outlined in Sect. 5.3.

5.1 Sequential Solution

Seaside operations planning basically comprises a single optimization problem regarding the service of vessels under limited quay space and QC capacity where the objective is to maximize an appropriate quality measure for the provided service. However, in practice as well as in the scientific literature, the complexity of this overall problem is broken down into subproblems of manageable complexity, namely the BAP, the QCAP, and the QCSP. The separate consideration of these problems calls for a hierarchy, which defines an order for solving them. The sequential solution process enables a clear distinction of responsibilities among involved planners and an unambiguous chronology of decision making. It has to be noted that hierarchical planning is by no means unique to terminal operations planning. It is also a well known concept in production planning. In the basic model of hierarchical production planning of Hax and Meal (1975), product items are aggregated to product families, which, in turn, are aggregated to product types. A sequential solution process decides first on the production program for product types, then on the lot sizes for product families, and, finally, on the lot sizes for the product items of a family. As typical for hierarchical planning, each particular decision has to respect thereby the decisions made at previous levels of the hierarchy.

The sequential solution process of the three CT seaside planning problems is sketched in Fig. 5.1 together with the relevant input and output data of each

F. Meisel, *Seaside Operations Planning in Container Terminals*,
Contributions to Management Science,
DOI: 10.1007/978-3-7908-2191-8_5, © Physica-Verlag Berlin Heidelberg 2009

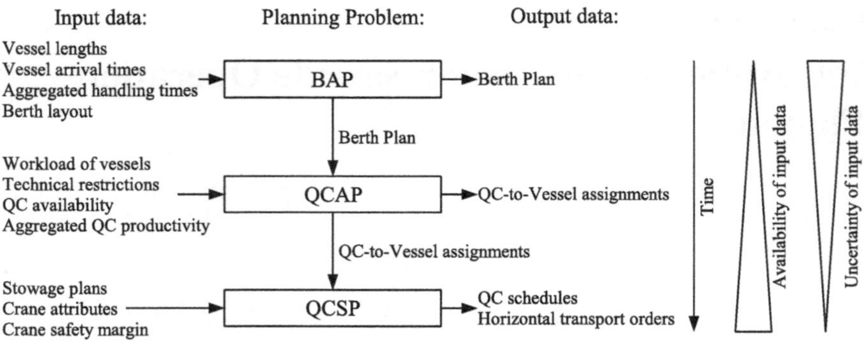

Fig. 5.1 Sequential planning of seaside operations

individual problem. As can be seen, the output (solution) of one problem may serve as a work plan for the terminal resources as well as an input for the subsequent planning problem.

Within the stated problem order the BAP is solved first. Vessel and quay data serve as a major input. Note that the handling time of a vessel represents aggregated data because the particular service process has not been planned yet. In the following QCAP, cranes are assigned to vessels with respect to the berthing times and berthing positions of vessels as derived within the BAP, i.e., the berth plan serves as an input for the crane assignment. Further input data is the workload of each vessel, e.g., the number of containers to charge and discharge, technical restrictions such as the maximum number of cranes to assign, and the availability of QCs within the planning horizon. Additionally, a QC productivity estimate can be used to decide on the crane capacity to assign to a vessel. At this stage, however, only empirical data, such as the number of moves per hour and crane observed at the terminal on average, can be used. Afterwards, the QCSP is solved, where the QC-to-Vessel assignments serve as an input. These assignments specify the availability of cranes at a vessel. Additionally, precise stowage plans are required for the detailed crane scheduling in order to derive the distribution of workload over the bays of a vessel.

The described sequential solution process reflects the increasing availability of input data and its decreasing uncertainty in the course of time. Nevertheless, the weakness of sequential planning becomes obvious in the light of the fact that the solution of a problem is based heavily on estimated input data. The BAP incorporates aggregated handling times while more precise handling times are obtained when the QCAP is solved afterwards. Similarly, the QCAP uses estimated QC productivity information although the realizable productivity of assigned cranes is revealed in the QCSP, specifying the QC capacity demand of a vessel. In the sequential solution process, decisions made at previous stages cannot be revised, even if the outcome of a subsequent planning problem does not fit the estimates previously used. In practice this leads to *ad-hoc modifications of plans* during their execution, whenever infeasibilities or poor performance are identified. If, for example, due to an insufficient assignment of crane capacity, the service of a vessel takes longer than expected, subsequently served vessels must be delayed or reassigned to other quay positions. Obviously, such modifications disturb the operations of the CT. In the best case, they cause idle times of quay space and cranes. More worse, ad-hoc modifications of berthing times and increased vessel handling times reduce the reliability of terminal services and thus, reduce the satisfaction of CT customers.

Enhanced seaside operations planning is based on *precise vessel handling times*, which are achieved through considering the QC resource within the berth planning. This requires a turn away from the problem hierarchy towards alternative integration concepts for seaside operations planning. To provide a framework for distinguishing different concepts, the first of two sequentially solved problems is referred to as the top-level problem in the following and the second is called the base-level problem, see Schneeweiss (2003). The base-level has to respect decisions made at the top-level. They are propagated in the form of instructions to the base-level. In seaside operations planning the BAP is a top-level problem for the QCAP which plays the role of a base-level problem. Moreover, the QCAP is the top-level problem to the QCSP.

According to Geoffrion (1999) integration of two problems can be done either by a *deep integration* or by a *functional integration*. Similar concepts are proposed by Muhanna and Pick (1988) and Dolk and Kottemann (1993) under different terms. Deep integration merges the top-level problem and the base-level problem into a single monolithic problem formulation, which makes a propagation of instructions obsolete. While solutions may be excluded from the search in a sequential solution process because of the incomplete consideration of the subproblem interrelations, deep integration enables to search the complete solution space of the overall problem. However, a deep integration causes in general a strong increase in the computational effort compared to a sequential solution process.

A functional integration is based on the original formulations of the problems. The integration is realized by a computational agenda that defines the sequence of the problems in the solution process and the data to exchange between the base-level and the top-level. Basically, functional integration of two problems can follow two possible ways. If the order given by the hierarchy is preserved, integration takes place by feeding back the output of the base-level as a further input for the top-level, see Fig. 5.2a. The top-level decisions are revised by solving the problem

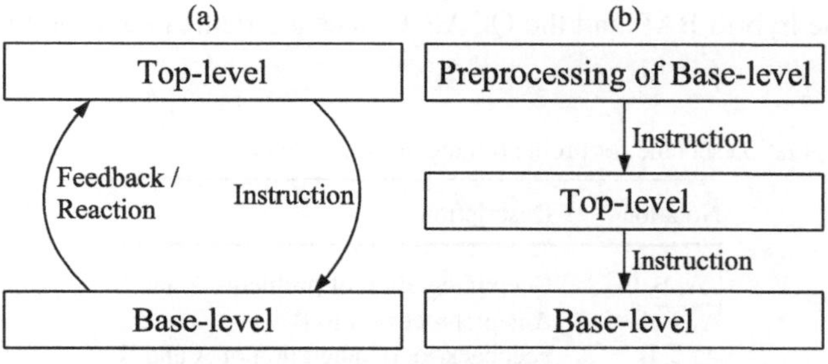

Fig. 5.2 Functional integration by a feedback loop (**a**) and by a preprocessing (**b**)

again using the feedback information. Such a *feedback loop* is performed iteratively until a certain termination criterion is met, e.g., a steady state is reached where no change in the solutions of the top-level problem is observed. The second way for a functional integration is to change the order of solving the problems, see Fig. 5.2b. Here, the original base-level problem is solved in a *preprocessing* phase to generate more detailed input data for the top-level problem. The success of this type of functional integration depends on the quality of base-level solutions. Note that these solutions have to be generated without knowing a top-level problem solution. The top-level problem incorporates information obtained in the preprocessing. Moreover, as shown in Fig. 5.2b, the base-level solutions of the preprocessing phase can be revised by finally solving the base-level problem again.

5.2 Integration Concepts in the Literature

Recent integration approaches for seaside operations planning motivate a further classification scheme, based on the concepts briefly introduced in the previous section. To distinguish problem integration by monolithic models (deep integration), by problem preprocessing, and by feedback loops, the notation of Table 5.1 is used. In this table, capitals A and B stand proxy for a BAP, QCAP, or QCSP. If a planning problem involves multiple decision variables but not all of them are determined at once in the integration model, the addressed decisions appear in brackets. For example, $\boxed{\text{BAP, QCAP(number)}}$ → QCAP(specific) stands for an integration where a berthing time and position, and a number of cranes are assigned to every vessel in a monolithic model, while the used cranes are specified subsequently.

Table 5.2 gives an overview of approaches for integrated seaside operations planning. In the following the mentioned papers are reviewed with respect to the used integration models. Approaches that exclusively use functional integration are proposed by Lee et al. (2006) and Lokuge and Alahakoon (2007). Lee et al. (2006) consider a feedback loop integration between the discrete BAP and the QCSP. There is no QCAP involved because the berths possess dedicated cranes. The solution of the BAP delivers a sequence for serving the vessels at each berth. A resulting QCSP is solved for every vessel and the obtained handling times are returned to the berth planning level to revise the vessel sequences. This loop is executed for a preset number of iterations. In Lokuge and Alahakoon (2007) a MAS is used to integrate the hybrid BAP and the QCAP. Cranes are dedicated to berths which are

Table 5.1 A notation scheme for problem integration concepts

Notation	Description
$\boxed{\text{A, B}}$	Deep integration of problems A and B
A → B	A is preprocessed to B
A ⇄ B	Feedback loop integration of A and B

Table 5.2 Overview of integration concepts for seaside operations planning

Integration concept	Reference
BAP ⇄ QCSP	Lee et al. (2006)
BAP ⇄ QCAP	Lokuge and Alahakoon (2007)
[BAP, QCAP(number)]	Oğuz et al. (2004), Meisel and Bierwirth (2006), Hendriks et al. (2008), Giallombardo et al. (2008), Liang et al. (2009)
[BAP, QCAP(number)] → QCAP(specific)	Park and Kim (2003)
[BAP, QCAP(number)] ⇄ QCAP(specific)	Imai et al. (2008a)
[BAP, QCAP]	Rashidi (2006), Theofanis et al. (2007b)
BAP ⇄ [QCSP, QCAP]	Meier and Schumann (2007)
[BAP, QCAP(number), QCSP] → QCAP(specific)	Ak and Erera (2006)
QCSP → [BAP(berthing times), QCAP]	Liu et al. (2006)
[QCAP, QCSP]	Daganzo (1989), Peterkofsky and Daganzo (1990), Tavakkoli-Moghaddam et al. (2009)

shared by vessels served at the same time. The problem is solved by software agents responsible for berth planning and communicating with other agents responsible for the crane assignment. The architecture of the used MAS constitutes a feedback loop integration of the BAP and the QCAP.

A deep integration of the continuous BAP and the QCAP is investigated by Oğuz et al. (2004), Meisel and Bierwirth (2006), Hendriks et al. (2008), Park and Kim (2003), Rashidi (2006), and Theofanis et al. (2007b). These papers present optimization models to decide on the berthing time, the berthing position, and the number of cranes for each vessel. The same decisions are considered for the discrete BAP by Giallombardo et al. (2008), Liang et al. (2009), and Imai et al. (2008a). In four of these approaches, the specific cranes used for the service of vessels are additionally determined. For this purpose different integration concepts are applied. Park and Kim (2003) consider the specific crane assignment as an end-of-pipe optimization which is appropriately solved in a postprocessing phase. In contrast, Imai et al. (2008a) return the specific crane assignment to the berth planning level where the made berthing decisions are evaluated and possibly revised. In two papers, the number and the specific set of cranes assigned to vessels are decided within a monolithic model. Rashidi (2006) merges the top-level problem and the end-of-pipe problem of Park and Kim (2003) into a single optimization model. A deep integration is also proposed by Theofanis et al. (2007b) for simultaneously assigning QCs and allocating vessels along the quay.

Several authors study integration models that involve all of the three seaside planning problems. A feedback loop integration of the BAP and the QCSP is described by Meier and Schumann (2007). The loop propagates a berth plan to the crane scheduling level. Detailed vessel handling times are obtained and returned to the top-level for an adjustment of the berth plan. The approach comprises a deep integration of QCAP and QCSP, because the crane schedules are collectively built for vessels served at the same time. Ak and Erera (2006) present an integration model that jointly decides on berth allocation, crane assignment, and crane scheduling. Merely the specific crane assignment is determined in a postprocessing, as has been proposed by Park and Kim (2003). The integration model of Liu et al. (2006) targets on the revision of a tentative berth plan. First, crane schedules are preprocessed to generate possible vessel handling times for each vessel and each assignable number of cranes. Next, specific cranes are assigned to vessels, where the tentative handling times are replaced by selecting values provided in the preprocessing phase. In order to minimize the maximum vessel tardiness, the tentative berthing times are revised in this model. The berthing positions are taken from the tentative berth plan.

Further integration models are formulated by Daganzo (1989) and Peterkofsky and Daganzo (1990). They combine the QCAP and the QCSP by simultaneously scheduling multiple cranes for a set of vessels. The authors remark that berthing decisions should be integrated with crane operations planning and illustrate this issue by examples under the assumption of identical sized vessels. Also Tavakkoli-Moghaddam et al. (2009) deal with the integration of crane assignment and scheduling. In this work the QCSP model of Kim and Park (2004) is extended such that multiple vessels are considered in parallel.

5.3 Designing a Comprehensive Integration Concept

As shown in the previous section, concepts for the integration of BAP, QCAP, and QCSP within the overall problem of seaside operations planning have seldom been investigated in scientific literature. Merely Liu et al. (2006), Ak and Erera (2006), and Meier and Schumann (2007) provide studies concerning this matter. In the following a new integration concept is presented in order to contribute to this field of research. The overall objective is to derive a concept that enables to determine berthing positions, berthing times, crane assignments, and crane schedules for the vessels with respect to the interrelations of the decisions fields. The following questions must be answered for the design of an integration concept:

1. Which variant of the BAP, the QCAP, and the QCSP is involved as subproblem in the overall problem of seaside operations planning?
2. How are the considered subproblems integrated within the overall problem?

The first question is answered by identifying problem characteristics that call for an integration. In CTs where the quay is not partitioned into berths, the assignment of cranes to vessels is most flexible and thus, the interrelations between BAP and QCAP are of particular importance. Hence, the continuous type of BAP is the most relevant problem variant for integrated seaside operations planning.

Moreover, respecting arrival times of vessels is indispensable for providing a satisfying service quality to vessel operators, calling for the consideration of the continuous dynamic BAP. A best possible assignment of cranes to vessels is enabled by considering variable-in-time QC-to-Vessel assignments. The important practical relevance of such assignments is revealed by the empirical investigation of Chu and Huang (2002). Hence, the crane assignment problem has to be formulated such that variable-in-time assignments are in its scope. The crane scheduling problem must be formulated at a reasonable level of detail in order to uncover the productivity loss caused by crane interference. A useful aggregation level for the QCSP is to define tasks by container groups. It allows cranes to share the workload of bays to a certain extent, while the computational effort is still moderate compared with scheduling single containers. The container group strategy preserves furthermore the grouping information of containers, which eases the planning of horizontal transport operations.

The question how to integrate the considered subproblems in the overall problem cannot be answered consistently. A deep integration of BAP and QCAP has been studied in diverse papers, proving that the resulting problem is still computationally tractable. Unfortunately, a deep integration of the QCSP into the BAP and/or the QCAP fails for practical reasons. In practice vessel operators have often not transmitted the stowage plans for vessels once the seaside operations are to be planned. Consequently, the required input data for the crane scheduling is not available. A functional integration is more flexible because it can be bypassed for vessels without available stowage plans. Hence, functional integration is useful for the integration of crane scheduling into berth planning and crane assignment.

The studies of Liu et al. (2006), Ak and Erera (2006), and Meier and Schumann (2007) consider the three seaside planning problems not on the level of abstraction described above. Liu et al. (2006) decide on the berthing times of the vessels, but assume that the berthing positions are given. Furthermore, variable-in-time QC-to-Vessel assignments are not in the scope of the approach. Also the approaches of Ak and Erera (2006) and Meier and Schumann (2007) show apparent weaknesses. For the proposed deep integration of the QCSP into the BAP or the QCAP, the relevant input data may not be available in practice. Furthermore, both studies define tasks on the basis of complete bays and ignore safety margins, which is inadequate for the integration of crane scheduling within the overall problem of seaside operations planning.

The new concept, which builds the basis for the integration of seaside planning problems in the thesis, is outlined in Fig. 5.3. It comprises a deep integration of BAP and QCAP within the *berth planning phase* and functional integrations of the QCSP in a *preprocessing phase* and a *feedback loop phase*. For the berth planning phase the concept of Park and Kim (2003) is taken up. It enables variable-in-time QC-to-Vessel assignments, and it decides also on the specific cranes that serve a vessel.

In the preprocessing phase, individual crane productivities for each vessel are generated in terms of crane utilization rates. This data is involved in the berth planning phase to generate appropriately dimensioned variable-in-time QC-to-Vessel assignments. To obtain precise crane productivities, a rich QCSP formulation has to

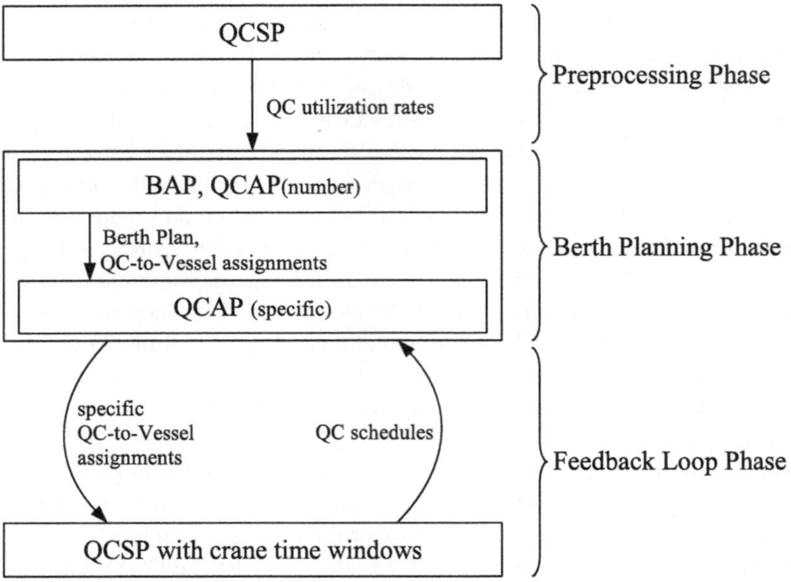

Fig. 5.3 A new concept for integrating seaside planning problems

be employed where tasks are defined by container groups and where safety margins, the non-crossing condition, and movement time of cranes are respected. Such a formulation is provided in the stream of research of Kim and Park (2004), Moccia et al. (2006), and Sammarra et al. (2007).

The feedback loop phase generates crane schedules for the QC-to-Vessel assignments derived in the berth planning phase. This requires an extension of the QCSP with respect to time windows for the cranes. Feeding back these crane schedules into the berth planning phase is necessary in order to adjust inappropriate crane assignments, berthing positions, and berthing times.

To summarize, the new integration concept represents the decisive interrelations between berth planning and crane operations planning. The subsequent chapters of the thesis provide studies that are concerned with modeling, solving, and linking of the optimization problems contained in the integration approach. Chapter 6 provides a study on the berth planning phase. In Chap. 7 crane scheduling is investigated as an isolated problem. Finally, in Chap. 8 the integration of crane scheduling into the berth planning phase is considered by investigating the preprocessing phase and the feedback loop phase in detail.

Chapter 6
Berth Allocation and Quay Crane Assignment

The new concept for integrated seaside operations planning comprises a deep integration of the BAP and the QCAP. The resulting problem, namely the *Berth Allocation and Crane Assignment Problem* (BACAP), is studied within this chapter. The first mathematical formulation of the combined problem of berth allocation and crane assignment has been presented by Park and Kim (2003). A new problem formulation has been provided by Meisel and Bierwirth (2009), which incorporates QC productivity determining effects. This new model is presented in Sect. 6.1 and solution methods are described in Sect. 6.2. Computational tests follow in Sect. 6.3. Section 6.4 concludes the BACAP study.

6.1 Modeling the BACAP

6.1.1 Problem Description and Assumptions

The BACAP bases on the continuous dynamic variant of the BAP. It is formally described as follows. A terminal with a quay of length L, measured in segments of 10 m length, is considered. A number of Q QCs is available to serve vessels. The planning horizon of the BACAP is H hours, where T is a corresponding set of 1-hour time periods, i.e., $T = \{0, 1, \ldots, H-1\}$. Within the planning horizon a set of vessels $V = \{1, 2, \ldots, n\}$ is projected to be served, where n is the total number of vessels.

For each vessel $i \in V$ its length l_i, measured in segments of 10 m length, is given. The crane capacity demand of vessel i to fulfill all loading and unloading operations is m_i QC-hours. The minimum and maximum number of QCs to assign to the vessel are denoted by $r_i^{\min}$ and $r_i^{\max}$, yielding the range $R_i = [r_i^{\min}, r_i^{\max}]$. Furthermore, an expected time of arrival ETA_i is given. Berthing the vessel earlier than ETA_i is possible by a speedup on its journey to the terminal. The realizable speedup, however, is bounded. To model this an earliest starting time $EST_i \leq ETA_i$ is given, i.e., the

F. Meisel, *Seaside Operations Planning in Container Terminals*,
Contributions to Management Science,
DOI: 10.1007/978-3-7908-2191-8_6,

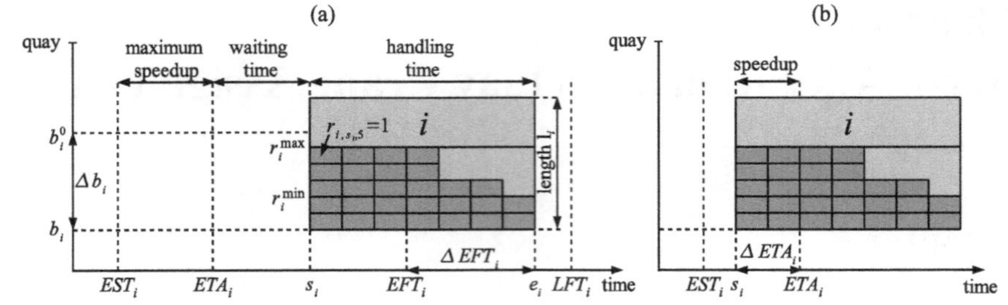

Fig. 6.1 Vessel data – waiting before berthing (**a**) and speedup case (**b**)

vessel cannot be berthed earlier than EST_i. Finally, an expected finishing time EFT_i and a latest finishing time LFT_i are given for the vessel. Import and export containers of a vessel are stored in dedicated yard areas. A desired berthing position b_i^0 is specified for vessel i within the vicinity of these yard areas.

The following assumptions are made for the BACAP:

1. Each quay position shows sufficient water depth to berth arbitrary vessels.
2. It takes no time to berth and to unberth vessels.
3. It takes no time to move a QC from one vessel to another vessel.
4. Vessels are served without preemption, i.e., once started to serve a vessel the process is not interrupted until the service is completed.
5. Every crane has the technical capability to serve every vessel. Furthermore, the cranes are identical, i.e., they show the same maximum productivity.

The decisions of the BACAP are to determine a berthing time s_i, a berthing position b_i, and the number of QCs to assign to each vessel $i \in V$ in its service periods such that a cost measure is minimized. The berthing time s_i of a vessel follows from the beginning of the first period with cranes assigned, whereas its departure time e_i is defined by the end of the last period with cranes assigned. The time span between s_i and e_i defines the handling time of vessel i. The assignment of cranes to vessels is represented by a binary decision variable r_{itq}. It is set to 1, if and only if exactly q QCs are assigned to vessel i at time t. To evaluate a solution to the BACAP, the deviation from the desired berthing position $\Delta b_i = |b_i^0 - b_i|$, the necessary speedup $\Delta ETA_i = (ETA_i - s_i)^+$, and the tardiness $\Delta EFT_i = (e_i - EFT_i)^+$ are determined for each vessel i. Figure 6.1 illustrates the interrelations of the so far introduced vessel data and variables. A description of the cost structure of a vessel follows in Sect. 6.1.3.

6.1.2 Resource Utilization

Different effects influence the productivity of a terminal and thus, the utilization of its resources. For seaside operations, two influencing factors are of importance and need to be incorporated in a BACAP formulation:

- Interference among QCs
- Berthing vessels apart from desired berthing positions

The rail mounted QCs in a CT are unable to pass each other. As a consequence *interference among QCs* can take place in the form of unproductive crane waiting time. In general, the more cranes are assigned to a vessel the more interference will take place leading to reduced marginal productivity of cranes. For reasons of simplicity Park and Kim (2003) ignore this effect by assuming that the crane productivity is proportional to the number of QCs that simultaneously serve a vessel. To overcome this simplification, crane productivity loss must be formally described. According to Schonfeld and Sharafeldien (1985) an *interference exponent* can be used that reduces the marginal productivity of cranes. For a given interference exponent α ($0 < \alpha \leq 1$), the productivity obtained from assigning q cranes to a vessel for one hour is given by a total of q^{α} QC-hours. This idea was taken up by Silberholz et al. (1991) to support the allocation of human resources in container terminals and by Dragovic et al. (2006) for a simulation study on the berthing process. Oğuz et al. (2004) transfer a similar idea from machine scheduling to berth allocation and crane assignment where the interference exponent is used to determine handling times of vessels instead of crane productivity. Unfortunately, the solution method adopted from machine scheduling considers time-invariant QC-to-Vessel assignments only. It furthermore necessitates the objective of makespan minimization, which is rarely considered in berth planning due to its low practical relevance.

The productivity of a terminal is also affected by the workload of horizontal transport means. This workload is minimal if a vessel berths at its *desired berthing position* b_i^0. If the actually chosen berthing position is apart from the desired position, the load of the horizontal transport increases. This effect can be partially alleviated by deploying more transport vehicles. Therefore, Park and Kim (2003) propose to penalize apart berthing positions through additional costs. The approach, however, ignores the fact that a larger number of vehicles decelerates the average speed and thus reduces the service rate again. Therefore an apart berthing position of a vessel leads to a productivity loss. This productivity loss is modeled by an increase in the vessel's QC capacity demand. Let $\beta \geq 0$ denote the relative increase of QC capacity demand per unit of berthing position deviation, called the berth deviation factor. Hence, a vessel positioned Δb_i quay segments away from its desired berthing position requires $(1+\beta\Delta b_i)m_i$ QC-hours to be served.

With respect to both effects described above, the minimum handling time needed to serve vessel i is given as

$$d_i^{\min} = \left\lceil \frac{(1+\beta\Delta b_i)m_i}{(r_i^{\max})^{\alpha}} \right\rceil. \tag{6.1}$$

As an example, let the handling of vessel i require a total of 15 QC-hours. The vessel can be served by at most five QCs simultaneously. Assume further that the vessel is berthed 100 m away from its desired position, which corresponds to $\Delta b_i = 10$ quay segments. Without productivity loss the fastest possible handling requires 3 hours. If the interference exponent and the berth deviation factor are set to $\alpha = 0.85$ and $\beta = 0.02$, respectively, the minimum handling time increases to 5 hours according to (6.1).

6.1.3 Cost Structure

The most frequently pursued objective in berth allocation models is the minimization of waiting and handling times of vessels in order to achieve a high satisfaction of vessel operators. For a precise treatment of the various factors influencing service quality, different cost functions are proposed in the literature, see, e.g., Park and Kim (2003), Golias et al. (2006), and Hansen et al. (2008). In the following, the *service quality cost* of vessel i is the sum of three types of cost:

- Speedup cost for catching a berthing time earlier than ETA_i
- Tardiness cost for exceeding the expected finishing time EFT_i
- Penalty cost for exceeding the latest allowed finishing time LFT_i

The corresponding cost rates are denoted as c_i^1, c_i^2, and c_i^3. While speedup cost and tardiness cost grow constantly in time, penalty cost incur only once, if the departure of the vessel is beyond the latest allowed finishing time LFT_i. Figure 6.2 illustrates the cost drivers of service quality on a discrete time basis. If the vessel is completely served in the time span between ETA_i and EFT_i, the perfect service quality is reached and no cost is incurred.

Service quality objectives are certainly of highest importance. Nevertheless, besides offering a competitive service, the CT management also has to pursue low operational costs. Regarding the seaside of a CT, one of the operational cost drivers is the labor force needed to operate the QCs. Therefore, Meisel and Bierwirth (2006) propose to minimize the number of 8-hour gang shifts required to fulfill a berth plan without considering any service objectives. To combine service quality objectives and resource cost objectives, a fourth cost type is added here, called the *QC operational cost*. It evaluates the utilized QC-hours within a berth plan. The objective accounts for the decreasing marginal productivity of QCs and the resulting trade-off between accelerating the handling of a vessel and the operational cost of QCs. The cost rate per QC-hour is denoted as c^4. The QC operational cost plus the service quality costs of the vessels make up the total service costs of a berth plan. In the following all cost rates are given in units of 1,000 USD.

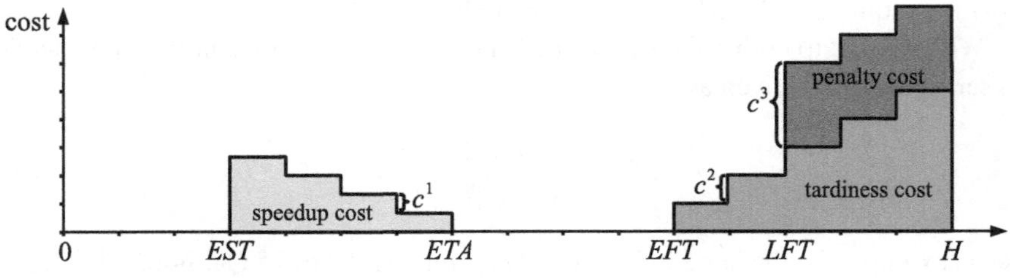

Fig. 6.2 Structure of the service quality cost of a vessel

6.1.4 Optimization Model

Besides the already introduced decision variables s_i, b_i, and r_{itq} the following binary decision variables are denoted to provide a mathematical formulation of the BACAP:

r_{it} Set to 1 if at least one QC is assigned to vessel i at time t, 0 otherwise
u_i Set to 1 if the finishing time of vessel i exceeds LFT_i, 0 otherwise
y_{ij} Set to 1 if vessel i is berthed below of vessel j, i.e., $b_i + l_i \leq b_j$, 0 otherwise
z_{ij} Set to 1 if the service of vessel i ends not later than the service of vessel j starts, 0 otherwise

The BACAP is formulated as follows:

$$\text{minimize } Z = \sum_{i \in V} \left[c_i^1 \, \Delta ETA_i + c_i^2 \Delta EFT_i + c_i^3 u_i + c^4 \sum_{t \in T} \sum_{q \in R_i} (q \, r_{itq}) \right] \quad (6.2)$$

subject to

$$\sum_{t \in T} \sum_{q \in R_i} (q^{\alpha} r_{itq}) \geq (1 + \beta \Delta b_i) m_i \qquad \forall i \in V, \quad (6.3)$$

$$\sum_{i \in V} \sum_{q \in R_i} (q r_{itq}) \leq Q \qquad \forall t \in T, \quad (6.4)$$

$$\sum_{q \in R_i} r_{itq} = r_{it} \qquad \forall i \in V, \forall t \in T, \quad (6.5)$$

$$\sum_{t \in T} r_{it} = e_i - s_i \qquad \forall i \in V, \quad (6.6)$$

$$(t+1) r_{it} \leq e_i \qquad \forall i \in V, \forall t \in T, \quad (6.7)$$

$$t r_{it} + H(1 - r_{it}) \geq s_i \qquad \forall i \in V, \forall t \in T, \quad (6.8)$$

$$\Delta b_i \geq b_i - b_i^0 \qquad \forall i \in V, \quad (6.9)$$

$$\Delta b_i \geq b_i^0 - b_i \qquad \forall i \in V, \quad (6.10)$$

$$\Delta ETA_i \geq ETA_i - s_i \qquad \forall i \in V, \quad (6.11)$$

$$\Delta EFT_i \geq e_i - EFT_i \qquad \forall i \in V, \quad (6.12)$$

$$M u_i \geq e_i - LFT_i \qquad \forall i \in V, \quad (6.13)$$

$$b_j + M(1 - y_{ij}) \geq b_i + l_i \qquad \forall i, j \in V, i \neq j, \quad (6.14)$$

$$s_j + M(1 - z_{ij}) \geq e_i \qquad \forall i, j \in V, i \neq j, \quad (6.15)$$

$$y_{ij} + y_{ji} + z_{ij} + z_{ji} \geq 1 \qquad \forall i, j \in V, i \neq j, \quad (6.16)$$

$$s_i, e_i \in \{EST_i, \ldots H\} \qquad \forall i \in V, \quad (6.17)$$

$$b_i \in \{0, 1, \ldots L - l_i\} \qquad \forall i \in V, \quad (6.18)$$

$$\Delta ETA_i, \Delta EFT_i \geq 0 \qquad \forall i \in V, \quad (6.19)$$

$$r_{itq}, r_{it}, u_i, y_{ij}, z_{ij} \in \{0, 1\} \qquad \forall i, j \in V, \forall t \in T, \forall q \in R_i. \quad (6.20)$$

This optimization model pursues the minimization of the total cost arising from the service of all vessels within the planning horizon. Constraints (6.3) ensure that every vessel receives the required QC capacity with respect to productivity losses by QC interference and the chosen berthing position. Note that the number of cranes q assigned to a vessel is no decision variable in order to ensure the linearity of this constraint. Instead, the number of cranes assigned to a vessel is described by binary variables r_{itq}, indicating whether exactly q QCs are assigned to vessel i at time t. Constraints (6.4) enforce that at most Q cranes are utilized in a period. In every period a certain number of QCs is assigned to every vessel, which is either zero or taken from the range R_i. A consistent setting of the corresponding variables r_{it} and r_{itq} is ensured by (6.5). Constraints (6.6)–(6.8) set the starting times and ending times for serving vessels without preemption. Constraints (6.9)–(6.12) determine the deviations from the desired berthing position, expected arrival time, and expected finishing time for each vessel. Variable u_i indicates whether the handling of vessel i ends later than LFT_i. It is set by Constraints (6.13) where M denotes a large positive number. Constraints (6.14) and (6.15) set the variables y_{ij} and z_{ij}, which are used to avoid overlapping the handling of vessels in the space–time diagram in Constraints (6.16). The arrival of a vessel can be sped up to at most the earliest starting time EST_i. Moreover, the planning horizon H defines a limit on the departure time of the vessels. Both aspects are reflected in Constraints (6.17). Constraints (6.18) ensure that each vessel is positioned within the quay boundaries. The Constraints (6.19) and (6.20) define domains for the remaining decision variables.

With the above model a linear formulation for the BACAP is provided. Although the BACAP model incorporates the productivity effects of resources, it is formulated more compact than the model of Park and Kim (2003) shown in Appendix A. The number of constraints grows in $O(nH)$ in the BACAP model while it grows in $O(nH^2)$ in the model provided by Park and Kim (2003). Also the number of variables grows less fast if Q is supposed to be much smaller than L. The compactness is based on a suitable formulation of the non-preemption condition in Constraints (6.6)–(6.8) and of the space–time condition in Constraints (6.14)–(6.16). The BACAP model shows that the assumption that QC productivity grows linearly in the number of cranes assigned to a vessel, as used by Park and Kim, can be replaced by a more accurate handling of crane productivity without increasing the complexity of the model.

The following characteristics of the classification scheme of Sect. 4.1.1 apply for the presented BACAP model. For the spatial attribute the value *cont* applies because the model is based on the continuous variant of the BAP. Since speeding up vessels is allowed only within certain bounds, an earliest possible time of arrival is known for every vessel. Due to the dynamic arrival process of vessels, the problem is classified as dynamic. The handling times of vessels depend on their berthing positions and the assignment of QCs as represented by the handling time characteristics *pos* and *QCAP*. The objective is the minimization of service quality costs incurred by speedups of vessels (*speed*) and by tardy departures (*tard*) as well as the operational cost of the utilized QC-hours (*res*). Note that the presented classification scheme does not distinguish between different types of tardiness costs for the sake of clarity. Summarizing, the BACAP is classified by $cont \mid dyn \mid pos, QCAP \mid \sum(w_1\, speed + w_2\, tard + w_3\, res)$.

The provided model can be reformulated to feature other well known BAP characteristics. For instance, discrete and hybrid BACAPs as well as vessel draft consideration can be modeled by eliminating forbidden berthing positions from the domains of the variables b_i. Due dates for the vessels can be represented in turn by an unacceptably large penalty c_i^3 for a tardy departure. The consideration of QC scheduling data within the BACAP, which leads to the handling time characteristic *QCSP*, is investigated by an explicit study later in this thesis.

6.2 Solution Methods

The BACAP as stated by (6.2)–(6.20) is an intractable problem because already the BAP is $\mathcal{NP}$-hard, see, e.g., Lim (1998) and Imai et al. (2005). Therefore, several heuristic solution methods are provided for the BACAP by Meisel and Bierwirth (2009):

- A construction heuristic to obtain an initial feasible solution
- Procedures for locally refining solutions by resource leveling and by shifting of vessel clusters
- Two meta-heuristics, namely Squeaky Wheel Optimization and Tabu Search

6.2.1 Construction Heuristic

To obtain an initial solution for this problem, a straightforward construction heuristic is used. It schedules the vessels one by one in the order of a given *priority list*. Vessel i is inserted into the partial berth plan by assigning it a berthing time s_i, a berthing position b_i, and the number q of cranes deployed in period t (represented by the variables r_{itq}). As shown in Fig. 6.3, the procedure *Insert(i)* performs eight steps, namely (a)–(h).

In Step (a) the cost for inserting vessel i, denoted here by Z_i^*, is initially set to infinity. In Steps (b) and (c) the berthing time for vessel i is set to the ideal berthing time ETA_i and the berthing position is set to the desired berthing position b_i^0.

In Step (d) an assignment of QCs is generated for the current position (s_i, b_i) in the space–time diagram by pursuing the fastest possible handling of the vessel. Using (6.1) the handling time $d_i^{\min}$ is computed leading to the ending time e_i. If the available number of QC-hours within this interval is insufficient to serve the vessel, respecting that no more than $r_i^{\max}$ QCs can be assigned to it within a period, e_i is increased until the capacity is sufficient. If either $e_i > H$ is observed or less than $r_i^{\min}$ QCs are available within at least one of the periods $[s_i, s_i+1, \ldots, e_i - 1]$ the QC assignment fails. Otherwise, a feasible QC assignment is obtained by assigning the available QCs within the determined handling interval respecting $r_i^{\min}$ and $r_i^{\max}$ until Constraint (6.3) holds for the vessel to be inserted. To minimize the productivity loss, an almost uniform distribution of QCs over time is realized. Pseudocodes

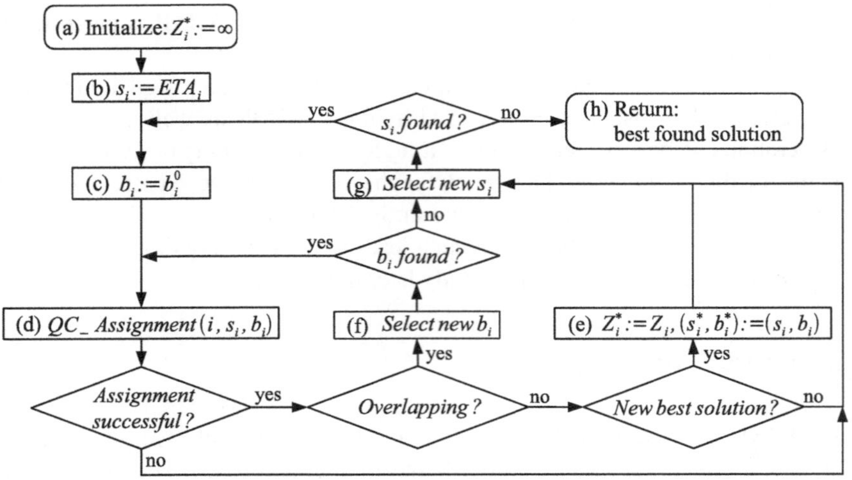

Fig. 6.3 Procedure *Insert(i)* of the constructor

for the crane assignment procedure and the subsequently described procedures are given in Appendix B.

If QCs have been assigned to a vessel, an ending time for its handling is fixed. To ensure consistency with the partial schedule, it is checked whether the vessel overlaps with other vessels in the space–time representation. If the attempted insertion is feasible, the cost of vessel i is computed according to the objective function (6.2). In the event that a new best solution has been found, its coordinates (s_i^*, b_i^*) and the corresponding cost Z_i^* are updated in Step (e). In case of an infeasible insertion, a new berthing position $b_i \in [0, L - l_i]$ is selected in Step (f). The new position is the closest not yet inspected position to the desired position b_i^0 such that the overlapping conflict is resolved. If such a position is found Step (d) is repeated. Otherwise, the procedure continues in Step (g), as it does if Step (d) has not delivered a successful QC assignment, or if a feasible schedule has been obtained.

In Step (g) a new starting time s_i for serving vessel i is taken one after the other from the list $[ETA_i - 1, ETA_i + 1, ETA_i - 2, \ldots]$ until s_i has reached EST_i in the one direction and the end of the planning horizon in the other. To speed up the procedure, a lower bound of the cost associated with a starting time is determined using (6.2). If the estimate overshoots the cost of the best known solution Z_i^* the iteration of earlier or later starting times is suppressed. The new starting time s_i is evaluated as described above. If no new starting time can be assigned to a vessel, the procedure *Insert(i)* terminates in Step (h), returning the best found solution (s_i^*, b_i^*).

Example 6.1: Insertion of a vessel

To illustrate the procedure, three vessels are assumed to be served at a terminal with $L = 14, H = 10, Q = 5, c^4 = 0.1, \alpha = 0.9$, and $\beta = 0.1$. The data of the vessels is shown in Table 6.1. The partial schedule, already fixed for Vessels 1 and 2, is

Table 6.1 Example vessel data

i	l_i	b_i^0	m_i	$r_i^{\min}$	$r_i^{\max}$	EST_i	ETA_i	EFT_i	LFT_i	c_i^1	c_i^2	c_i^3
1	3	7	4	1	3	2	3	5	8	1	1	2
2	4	7	10	1	2	2	3	9	10	2	2	4
3	5	6	5	1	3	1	4	6	7	3	3	6

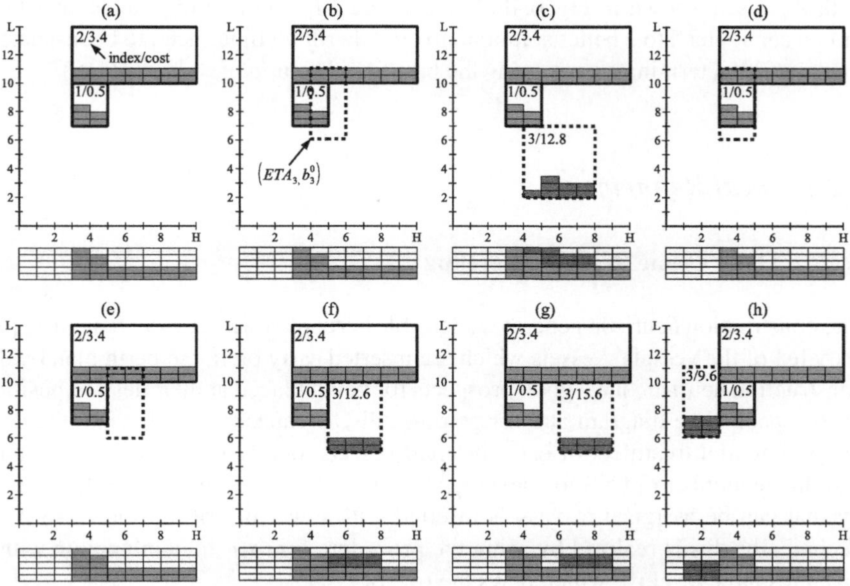

Fig. 6.4 Example positioning of a vessel

shown in Fig. 6.4a. Now, Vessel 3 has to be inserted. According to (6.1) its minimum handling time is $d_3^{\min} = 2$ hours if berthed at its desired berthing position.

At first, procedure $Insert(i = 3)$ selects the preferred coordinates, as shown by the dotted rectangular in Fig. 6.4b. Since this insertion is overlapping with the given partial schedule, Vessel 3 is repositioned to $b_3 = 2$. This berthing position deviates from the desired position by four quay segments. Therefore, the number of needed QC-hours increases from $m_3 = 5$ to $(1 + \beta \cdot 4)m_3 = 7$ QC-hours. As shown in Fig. 6.4c, the QC assignment procedure delivers $r_{3,4,1} = r_{3,5,3} = r_{3,6,2} = r_{3,7,2} = 1$, indicating that the number of assigned QCs changes twice within the service. This resource assignment is sufficient because the QC productivity of $1^{0.9} + 3^{0.9} + 2 \times 2^{0.9} = 7.42$ satisfies the needed 7 QC-hours. The projected service of the vessel requires no speedup ($\Delta ETA_3 = 0$), but the finishing time exceeds the expected finishing time ($\Delta EFT_3 = 2$) and also the latest allowed finishing time ($e_3 > LFT_3$). With 8 QC-hours assigned, the corresponding cost of the vessel is $Z_3 = 12.8$. Next, to generate an alternative berth plan, $ETA_3 - 1$ is assigned to the vessel as a new handling start time. Figure 6.4d shows that not enough QCs are available in this period with respect to $r_3^{\min}$. Therefore, the starting time is set to $s_3 = ETA_3 + 1$. As shown in Fig. 6.4e the insertion overlaps the partial berth plan again. This is resolved by repositioning the vessel to $b_3 = 5$ which enlarges the handling time of the vessel by one period leading to a new best solution with $Z_3 = 12.6$, see Fig. 6.4f. The generation of the next berth plan for $s_3 = 2$ fails due to short QC capacity in period 3. Continuing with $s_3 = 6$ delivers the solution shown in Fig. 6.4g. Next, $s_3 = 1$ is assigned to the vessel at its desired berthing position. This leads again to a feasible and new best solution with $Z_3 = 9.6$, shown in Fig. 6.4h. Since $s_3 = EST_3$ and berthing times later than time 6 cannot lead to a better solution, no other berthing times need to be inspected. The algorithm terminates, returning the best found solution $(s_3^*, b_3^*) = (1, 6)$.

6.2.2 Local Refinements

6.2.2.1 Quay Crane Resource Leveling

The construction heuristic generates a feasible berth plan with respect to a given priority list of the vessels. Vessels which are inserted early on in the berth plan by the construction heuristic have good prospects for being placed at their desired position in the space–time diagram and for getting full QC capacity. To alleviate the double preferential treatment of early inserted vessels, one can restrict the maximum assignable number of QCs to a level r_i^{lvl} below $r_i^{\max}$. This, in turn, saves QC capacity that can be assigned to vessels inserted with lower priority in the berth plan. Technically, this is realized by using the procedure *Insert(i)* in combination with a given resource level r_i^{lvl} which plays the role of $r_i^{\max}$.

The first refinement procedure considers the vessels one by one according to the given priority list $P = (p_1, p_2, \dots, p_n)$ of all vessels $i \in V$. Starting with an empty berth plan, vessel p_1 is inserted once for every resource level $r_{p_1}^{\text{lvl}}$ within the range R_{p_1}. Each of these incomplete berth plans is completed by subsequently inserting the remaining vessels p_2 to p_n using the insertion procedure without a restricting resource level. Due to the resource restriction for vessel p_1, other vessels have received higher priority regarding the QC assignment in the completed berth plans. Possibly, the saved QC capacity has not been completely exhausted by these vessels and can therefore be reassigned to vessel p_1. For this purpose, p_1 is removed again in all completed berth plans and inserted once again without a resource restriction. Afterwards, a partial berth plan containing vessel p_1 is obtained from the best of the generated solutions. Next, the partial berth plan of p_1 is extended by inserting vessel p_2 in the same manner. This process is continued for every vessel up to p_{n-1}. The vessel with the lowest priority is simply inserted in the almost completed berth plan with respect to the remaining QC capacity.

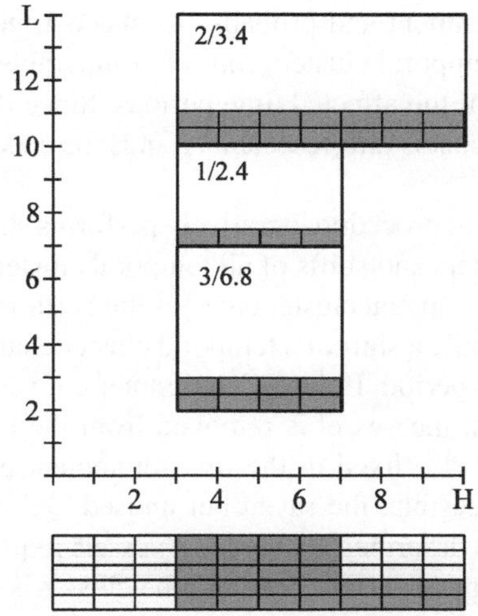

Fig. 6.5 Refinement of a berth plan by resource leveling

Example 6.2: Local refinement by resource leveling (continued Example 6.1)

To illustrate the procedure, a possible local refinement of the berth plan shown in Fig. 6.4h is described. If Vessel 1 is inserted with the resource level $r_1^{\text{lvl}} = 1$, accepting an increase in the handling time, Vessel 3 benefits from the saved QC capacity because its berthing time approaches its expected time of arrival. Vessel 2 is inserted as before leading to the improved berth plan shown in Fig. 6.5. While in this small example all vessels show a time-invariant QC-to-Vessel assignment, resource leveling also supports changes in the number of assigned QCs.

6.2.2.2 Spatial and Temporal Shifts

A further refinement aims at reducing cost by shifting clusters of vessels in the space–time diagram. According to Kim and Moon (2003) a spatial cluster is a subset of vessels that are connected in the space–time diagram because they occupy adjacent quay segments and are served simultaneously for at least one time period. A temporal cluster is a subset of vessels that are connected because they are served immediately one after the other, where subsequent vessels occupy at least one common quay segment.

In the approach of Kim and Moon (2003) the sets of spatial and temporal clusters are identified for a given berth plan. Afterwards, spatial clusters are shifted in the spatial dimension and temporal clusters in the temporal dimension, each as long as no further cost reduction is reachable. A similar concept is applied in Imai et al. (2005) where two conflicting vessels are shifted together like a single vessel. Both approaches do not take the QC assignment into consideration and require adaptation to be used for the BACAP. Since a spatial shift of a vessel changes its QC capacity demand, its crane assignment and probably its space–time positioning have to be revised. Shifting of a temporal cluster requires comparable revisions to respect the available QC capacity of the affected time periods. Since the impact of a spatial or a temporal shift on the cost is unforeseeable, it must be executed in order to identify improvements.

The second refinement procedure iteratively performs shifts of all spatial clusters towards the quay's borders and shifts of all temporal clusters within the entire planning interval. A shift of a spatial cluster changes the berthing position of each vessel by one quay segment while a shift of a temporal cluster changes the berthing time of each vessel by one time period. If the QC assignment of vessel i becomes infeasible due to a shift operation, the vessel is removed from the berth plan and reinserted with the resource level r_i^{lvl}, fixed in the first refinement phase. If all vessels of a cluster are scheduled feasible, the saved but unused QC capacity is reassigned to the reinserted vessels as described above. If all vessels require reinsertion, the structure of the cluster is supposed to be lost and the cluster is shifted no further in the considered direction. Improved solutions are recorded during the second refinement phase. It terminates if no further improvement is possible.

Example 6.3: Local refinement by vessel shifts (continued Example 6.2)

To illustrate the procedure, the spatial cluster $\{1,2,3\}$ shown in Fig. 6.5 is shifted towards the lower quay border. The first shift yields the berth plan of Fig. 6.6a. Now, Vessel 2 requires less QC capacity while the demands of Vessels 1 and 3 increase because they are shifted away from their desired positions. The existing QC assignments become infeasible. In the following reinsertion, Vessel 3 is assigned an

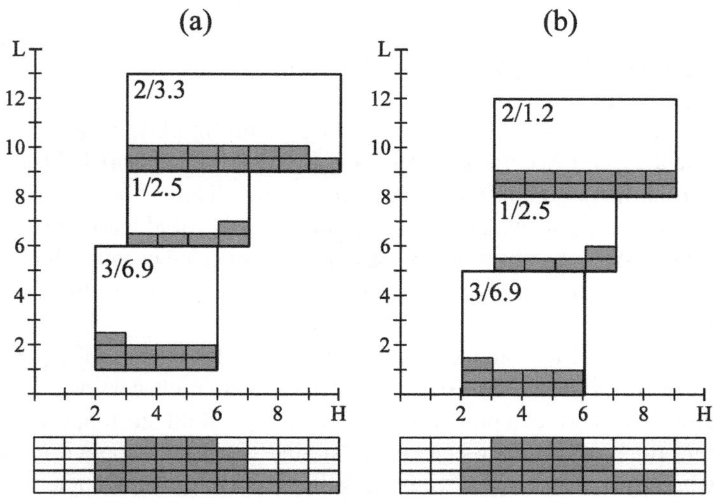

Fig. 6.6 Refinement of a berth plan by vessel shifts

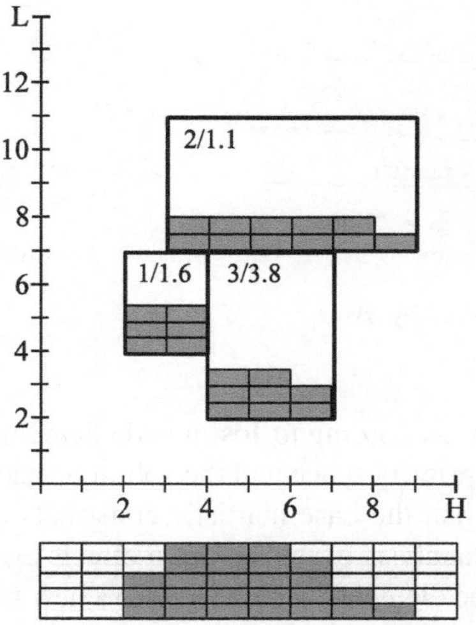

Fig. 6.7 Optimal solution to the BACAP example

earlier berthing time and Vessel 1 receives a released capacity unit, although its resource level r_1^{lvl} has been set to one by the previous refinement. With the next shift, the changed QC assignments of Vessels 1 and 3 are still feasible. The capacity demand of Vessel 2 decreases again and leads to cost 1.2 as shown in the improved berth plan, see Fig. 6.6b.

The optimal berth plan to the problem is shown in Fig. 6.7. This solution cannot be generated by the construction heuristic from the insertion order $(1,2,3)$ of the vessels. Hence, alternative priority lists for inserting vessels have to be taken into consideration.

6.2.3 *Meta-heuristics*

In this section two meta-heuristic approaches are presented, which enable changes in the priority list in order to improve the quality of berth plans.

6.2.3.1 Squeaky Wheel Optimization

Solutions of combinatorial optimization problems are often composed of elements with individual contributions to the overall solution quality. The idea of *Squeaky Wheel Optimization* (SWO), as introduced by Clements et al. (1997), is to exploit this information. In SWO a given solution is analyzed regarding the performance of its elements. In order to strengthen the overall performance, weak performing elements are assigned higher priority in the solution process by moving them towards the top of a priority list. The new list serves to build a new solution using a base

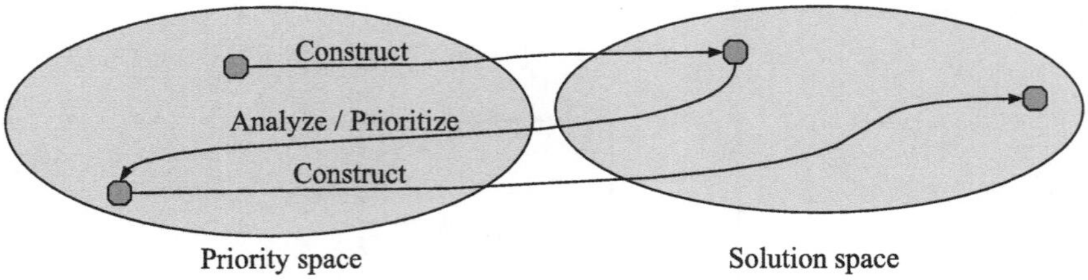

Fig. 6.8 Search spaces explored by SWO

heuristic of the problem. According to Joslin and Clements (1999), SWO searches two spaces, namely the priority space and the solution space, as shown in Fig. 6.8.

For a given priority list, the base heuristic constructs a corresponding point in the solution space. The analysis of this solution effects again a modification of the priorities of the contained elements, which leads to a new point in the priority space. The underlying strategy of SWO is to explore new solutions by large coherent moves in the priority space, which have only little chance to be reached through sequential moves in the solution space.

SWO has been used in a number of recent approaches to different combinatorial optimization problems, see Smith and Pyle (2004), Lim et al. (2004a,b), Fu et al. (2007). However, in these approaches SWO is rarely competitive to other meta-heuristics such as Genetic Algorithms and Tabu Search. It fails whenever the problems do not allow for a quantification of the individual contribution of each single problem element to the overall solution quality.

In the BACAP a berth plan (solution) is composed of vessels (elements) with individual cost contributing to the overall solution quality. The objective of the BACAP is to minimize the total service cost of the set of vessels. Therefore, SWO is straightforward applicable. Weak performing vessels are easily identified because they contribute relatively large proportions to the observed total cost. To generate new promising solutions, SWO increases the priority of these vessels at the expense of vessels with a lower service quality cost.

Initially, the priority list P of the vessels is ordered with respect to increasing arrival times. Ties are broken arbitrarily. The construction heuristic serves as a base heuristic in the SWO procedure to generate a berth plan for a given priority list. Hence, for the initial list, a berth plan is generated in a First-come First-served manner. Afterwards, a local refinement of the berth plan is done as described in Sect. 6.2.2 leading to an individual service quality cost for each single vessel. The operational costs for QCs are neglected in the solution analysis to avoid the bias that stems from the different QC capacity demand of vessels.

Following this solution analysis, the priorities of vessels are changed by a modification of the priority list. Two consecutive vessels in the priority list are swapped, if the cost incurred by the first vessel is lower than the cost incurred by the second vessel. Starting from the top, the priority list P is partially sorted by applying the swap operation $n-1$ times, which may lead to a multiple of changes. For a new priority list the corresponding berth plan is generated by the construction heuristic and the local refinements. Regardless of its quality, the obtained berth plan is accepted as a new solution and the SWO procedure starts a new round by analyzing the new solution. This time, the priorities are changed according to each vessel's total service quality cost of the first and the second solution. Doing so, vessels are prioritized according to their performance in all solutions generated so far.

SWO may be trapped in a cycle. Usually, this takes place if it generates a priority list that has already been generated in a previous round. A major source of cycling is local refinement, which may effect that a changed priority list leads to an already investigated solution and, thus, the priority list itself is also not influenced by this solution. Therefore, if a cycle is detected, the local refinement procedures are deactivated in SWO. The berth plans generated next will show worse quality and lead to changes in the priority list. Local refinements are reactivated if a not yet investigated priority list is found. The SWO procedure terminates after analyzing a given number of solutions without finding a new best solution.

Example 6.4: Prioritization by SWO (continued Example 6.3)

As an example for an iteration of SWO, the solution obtained after the second local refinement is taken up, see Fig. 6.6b. The original priority list that led to this solution is $P = (1,2,3)$. Since SWO considers service quality cost only, the relevant costs of Vessels 1, 2, and 3 are 2, 0, and 6, respectively. The priority list is changed by pairwise comparison of vessels on the basis of these costs. Hence, Vessels 1 and 2 do not change their position within the priority list because Vessel 1 shows higher service quality cost. The positions of Vessels 2 and 3 are changed because Vessel 3 shows higher service quality cost. A new priority list $P = (1,3,2)$ is derived, which is investigated in the next SWO iteration, and so on.

6.2.3.2 Tabu Search

As a further meta-heuristic approach the well known *Tabu Search* (TS) method, see Glover (1986), is applied to the BACAP. Like SWO, the proposed TS algorithm works on the priority list P of the vessels. Contrasting SWO, TS employs pairwise-exchanges of vessels in the priority list to obtain new solutions instead of adjacent swaps. The pairwise-exchange neighborhood of a solution is completely explored within each TS iteration. Every neighbor of the current solution, i.e., every modified priority list, is evaluated by the construction heuristic. If the obtained berth plan is an element of the tabu list, it is not considered any further. In order to save computation time, a local refinement is carried out only for the best performing neighbor of the current solution. This solution replaces the current solution even if it shows larger cost. The tabu list is managed as follows. The current berth plan is stored without the local refinement in the tabu list. In doing so, the totality of priority lists leading to this berth plan is set tabu at one strike. Since, by this effect, TS cannot benefit from removing any berth plan from the tabu list again, all berth plans are kept within the tabu list throughout the solution process. Using an Aspiration Criterion is not necessary because a new best solution found cannot be contained in the tabu list.

The TS algorithm terminates after a given number of iterations without finding a new best solution.

6.2.4 Specific Quay Crane Assignment

As intended by the provided BACAP model, the solution methods presented so far decide on the berthing times, the berthing positions, and the number of cranes to assign to each vessel within each period of its service interval. So far undecided is the set of *specific* cranes that make up the assigned QCs, see Fig. 3.4 on page 22 for an example. The determination of a specific assignment leads to a subsequent problem as shown by Park and Kim (2003), Ak and Erera (2006), and Imai et al. (2008a). Park and Kim (2003) propose a dynamic programming method to solve this problem. The method minimizes the total number of QC setups at the vessels and ensures that the cranes do not cross. It can be applied to the above BACAP formulation without modifications. The method is computationally inexpensive. For example, it generates a specific QC assignment for a BACAP solution with 40 vessels in less than a second. Therefore, the method is not involved in the computational study.

6.3 Computational Study

The following tests assess the performance of the BACAP solution methods and investigate the sensitivity of the solutions regarding the parameter settings.

- Performance comparison of BACAP solution methods:

  Test 6.1: Capability of CPLEX to deliver optimal solutions
  Test 6.2: Comparison of initial solutions and locally refined solutions
  Test 6.3: Comparison of SWO and TS
  Test 6.4: Comparison with the Park–Kim approach
  Test 6.5: Comparison with a sequential solution approach

- Sensitivity on problem parameter settings:

  Test 6.6: Effectiveness of vessel priorities
  Test 6.7: Estimating cost of productivity losses
  Test 6.8: Effectiveness of QC operational cost consideration
  Test 6.9: Potential of variable-in-time QC-to-Vessel assignments

In order to carry out the tests, all solution methods have been implemented in JAVA. A PC P4 2.4 GHz is used for the computations.

Table 6.2 Technical specifications and cost rates for different vessel classes

Class	l_i	m_i	$r_i^{\min}$	$r_i^{\max}$	c_i^1	c_i^2	c_i^3
Feeder	$U[\ 8,21]$	$U[\ 5,15]$	1	2	1	1	3
Medium	$U[21,30]$	$U[15,50]$	2	4	2	2	6
Jumbo	$U[30,40]$	$U[50,65]$	4	6	3	3	9

Benchmark Instances

For the tests, appropriate benchmark instances are required. While a set of benchmarks is provided by Park and Kim (2003) the instances are not rich enough to investigate all objectives of the outlined tests. For example, all vessels within these instances show identical cost rates. For this reason the instances are applied only in the comparison in Test 6.4. For all other tests a set of new created test instances is used. In these instances vessels are distinguished by three classes, namely feeder, medium, and jumbo. The classes differ in technical specifications and cost rates as shown in Table 6.2, where U expresses a uniform distribution of integer values in the specified interval. The given ranges are in accordance with empirical data provided by ISL (2003).

Three sets of test instances have been generated containing 20, 30, and 40 vessels with ten instances each. Within each instance, 60% of the vessels belong to the feeder class, 30% belong to the medium class, and 10% belong to the jumbo class. The planning horizon H is set to 1 week (168 h). The expected times of arrival ETA_i of vessels are uniformly distributed in the planning horizon. It is assumed that a vessel can speed up at most 10% which determines the earliest starting time $EST_i = \lceil 0.9ETA_i \rceil$. The expected finishing time EFT_i is derived by adding a vessel's minimum handling time to ETA_i. The latest finishing time LFT_i is derived by adding 1.5 times a vessel's minimum handling time to ETA_i. Further model parameters are as follows. The terminal data is $L = 100$ (1,000 m), $Q = 10$ QCs, and $c^4 = 0.1$ thousand USD per QC-hour. The desired berthing position is drawn for vessel i using $U[0, L - l_i]$. To attain moderate QC productivity losses, the interference exponent is set to $\alpha = 0.9$ and the berth deviation factor is set to $\beta = 0.01$. The latter effects a 1% increase in the handling effort per quay segment of berthing position deviation.

Since the planning horizon H imposes a hard constraint in the proposed BACAP model, the generated instances are not necessarily solvable. To ensure solvability, it is checked for every generated instance whether the construction heuristic returns a feasible solution. Only in this case the instance is included in an instance set.

Test 6.1: Capability of CPLEX to deliver optimal solutions

To obtain insight into the difficulty of the three instance sets, ILOG CPLEX 9.1 is applied using the options "emphasize optimality" and "aggressive cut generation". For recommendations of CPLEX parameter settings see Atamtürk and Savelsbergh

Table 6.3 CPLEX results for the test instances

$n=20$			$n=30$			$n=40$		
#	*Z*	*LB*	#	*Z*	*LB*	#	*Z*	*LB*
1	84.1	84.0	11	–	137.7	21	–	165.7
2	53.9*	53.9	12	81.8	81.4	22	–	159.6
3	77.4	75.2	13	104.9	100.9	23	–	185.0
4	76.2	75.8	14	–	96.8	24	–	224.1
5	56.8*	56.8	15	–	136.9	25	–	133.3
6	57.6*	57.6	16	–	106.2	26	–	201.3
7	68.0	67.5	17	–	99.6	27	–	172.2
8	56.1*	56.1	18	–	117.8	28	–	211.7
9	75.1	75.0	19	–	156.4	29	–	180.3
10	90.9	88.2	20	–	125.6	30	–	170.1

*Optimal solution

(2005). Table 6.3 reports the objective function value *Z*, representing the total cost of a berth plan, for each of the 30 instances, if found within a limited runtime of 10 h. Additionally, a lower bound *LB* is obtained from the solver and reported in every case. CPLEX always delivers near optimal solutions for small sized instances with 20 vessels. Note that these instances represent situations with low workload in a CT. Merely four instances (#2, 5, 6, 8) were proven to be solved to optimality within the given runtime limit. Most of the medium-sized instances remain unsolved, while not a single integer feasible solution has been found for the large sized instances. Running CPLEX with the option "emphasize feasibility" did not lead to further feasible solutions. These results indicate that the more congestion is faced at a CT the poorer CPLEX performs. While CPLEX does not provide a suitable solution procedure for the BACAP, the derived lower bounds are valuable for evaluating heuristic solutions in the subsequent tests.

Test 6.2: Comparison of initial solutions and locally refined solutions

This test investigates the quality of solutions obtained by the construction heuristic and by the local refinements. With vessels sorted by increasing expected time of arrival, the construction heuristic is used in First-come First-served manner, referred to as FCFS. To assess the individual contribution of the two local refinement procedures, the FCFS solutions are refined once by applying QC resource leveling (FCFS_{LR1}) and once by shifting vessel clusters (FCFS_{LR2}). Finally, both refinement procedures are subsequently applied to the initial solutions (FCFS_{LR}), where QC resource leveling is performed before shifting vessel clusters. The reverse order is not investigated because refinements of a berth plan obtained by shifts of vessel clusters get lost by a subsequent QC resource leveling. Table 6.4 reports the obtained objective function value *Z* and the relative error *RE* in percent of the heuristics

Table 6.4 Initial solutions and locally refined solutions

n	#	FCFS		FCFS_{LR1}		FCFS_{LR2}		FCFS_{LR}	
		Z	RE	Z	RE	Z	RE	Z	RE
20	1	118.5	41.07	118.5	41.07	86.1	2.50	86.1	2.50
	2	60.1	11.50	60.1	11.50	53.9	0.00	53.9	0.00
	3	97.6	29.79	97.6	29.79	87.3	16.09	87.3	16.09
	4	96.4	27.18	96.4	27.18	79.7	5.15	79.7	5.15
	5	73.1	28.70	65.2	14.79	56.8	0.00	56.8	0.00
	6	57.6	0.00	57.6	0.00	57.6	0.00	57.6	0.00
	7	93.3	38.22	91.6	35.70	71.4	5.78	69.9	3.56
	8	78.9	40.64	69.6	24.06	66.5	18.54	69.6	24.06
	9	96.4	28.53	96.4	28.53	76.3	1.73	76.3	1.73
	10	115.5	30.95	109.7	24.38	98.2	11.34	101.1	14.63
30	11	216.0	56.86	187.5	36.17	148.6	7.92	152.6	10.82
	12	96.7	18.80	94.7	16.34	87.9	7.99	86.4	6.14
	13	135.0	33.80	135.0	33.80	107.6	6.64	107.6	6.64
	14	144.5	49.28	130.9	35.23	117.5	21.38	113.2	16.94
	15	197.5	44.27	181.3	32.43	174.1	27.17	173.8	26.95
	16	137.7	29.66	132.1	24.39	125.8	18.46	127.2	19.77
	17	139.8	40.36	130.8	31.33	106.3	6.73	110.2	10.64
	18	167.8	42.44	167.8	42.44	131.4	11.54	131.4	11.54
	19	268.7	71.80	268.7	71.80	185.0	18.29	185.0	18.29
	20	184.7	47.05	178.1	41.80	144.3	14.89	140.5	11.86
40	21	317.0	91.31	278.3	67.95	298.5	80.14	261.3	57.69
	22	276.9	73.50	247.0	54.76	186.9	17.11	189.0	18.42
	23	550.4	197.51	364.0	96.76	455.0	145.95	325.7	76.05
	24	453.3	102.28	430.9	92.28	367.9	64.17	360.2	60.73
	25	239.1	79.37	208.6	56.49	166.6	24.98	162.0	21.53
	26	398.9	98.16	375.9	86.74	295.7	46.90	273.1	35.67
	27	354.6	105.92	292.7	69.98	245.6	42.62	233.0	35.31
	28	424.2	100.38	424.2	100.38	408.5	92.96	408.5	92.96
	29	334.2	85.36	289.7	60.68	291.2	61.51	268.4	48.86
	30	425.8	150.32	364.2	114.11	327.3	92.42	280.8	65.08
ARE (%)			59.83		46.76		29.03		23.99

against the CPLEX lower bound, i.e., $RE = (Z - LB)/LB \times 100$. To compare the heuristics on an aggregate level, the observed relative error has been averaged over the 30 test instances (*ARE*). Since all procedures terminate within less than a second for each of the instances, no computation times are reported here.

In Table 6.4 it can be seen that the initial solutions show very large relative errors of about 60% on average. For the large-sized instances with $n = 40$, the solutions even show an average error of more than 100%. Applying the local refinement procedures individually reduces the *ARE* considerably. Local refinement through resource leveling decreases the *ARE* by more than 13% compared to FCFS. Local refinement by shifting of vessel clusters decreases the *ARE* by more than 30%. Obviously, the shifting of vessel clusters provides a more effective local refinement of the initial solutions. However, better solution quality is obtained if both refinement procedures are applied sequentially to the initial solutions. Here, the *ARE* is about 36% below the *ARE* of the initial solutions. Taking a closer look at the three instance sets, one can see that for the small-sized instances with $n = 20$ only one instance (#7) is further improved, while for two instances (#8, 10) even worse solutions are returned by the sequentially applied local refinements compared to FCFS_{LR2}. Four of the medium-sized instances are improved (#12, 14, 15, 20), while for three instances (#11, 16, 17) worse solutions are returned compared to FCFS_{LR2}. The major improvements are observed for the large-sized instances with $n = 40$. Here, the sequential application of both refinement procedures improves eight of the ten instances compared to FCFS_{LR2}. These results show that shifting of vessel clusters successfully preserves the refinements obtained by QC resource leveling as encoded in the resource level variables r_i^{lvl}. From a practical point of view it becomes obvious that a higher congestion at a CT calls for the application of both refinement procedures.

Test 6.3: Comparison of SWO and TS

In this test, the two meta-heuristics Squeaky Wheel Optimization (SWO) and Tabu Search (TS) are compared. The initial priority list for the heuristics is derived from sorting the vessels by increasing expected time of arrival, i.e., the FCFS rule is applied. Both algorithms terminate after 200 iterations without gaining an improvement. Further parameters do not exist for the methods. Table 6.5 reports the obtained objective function value Z, the relative error *RE* against the CPLEX lower bound, and the computation times *time* (in seconds) for each of the 30 instances and each of the two meta-heuristics. To ease identification of the improvements realized by the meta-heuristics, the locally refined initial solutions FCFS_{LR} as found within the previous test are reported again.

The results show that SWO and TS deliver much better solutions than the local refinements for the 30 instances. The *ARE* observed for FCFS_{LR} is decreased by about 11% using SWO and by about 9% using TS. For small-sized instances SWO and TS return better solutions for six and seven instances, respectively. Here, the TS solutions for instances #4, 8, 9, 10 are better than the SWO solutions. However, for the instances with $n = 30$ and $n = 40$ SWO is superior to TS. For example, for the large-sized instances with $n = 40$, SWO shows an *ARE* of about 30% but TS shows an *ARE* of 33%. The increase in the *ARE* for medium-sized and large-sized instances shows that the BACAP becomes more difficult to solve if the workload increases in the terminal. This is also reflected by the growth of average cost per vessel under an increasing workload. For instances with 20 vessels, average cost of approximately 3,500 USD per vessel are observed. They increase to 5,900 USD per vessel for instances with 40 vessels.

Regarding the runtimes, SWO is slightly faster than TS within each of the three instance sets. The average runtime per instance significantly increases from smaller to larger instances for both meta-heuristics but stays clearly below 10 min even for

Table 6.5 Performance comparison of meta-heuristics

n	#	$FCFS_{LR}$		SWO			TS		
		Z	RE	Z	RE	$time$	Z	RE	$time$
20	1	86.1	2.50	85.1	1.31	11	85.1	1.31	14
	2	53.9	0.00	53.9	0.00	4	53.9	0.00	8
	3	87.3	16.09	77.4	2.93	11	77.4	2.93	17
	4	79.7	5.15	79.7	5.15	8	77.9	2.77	12
	5	56.8	0.00	56.8	0.00	10	56.8	0.00	15
	6	57.6	0.00	57.6	0.00	4	57.6	0.00	7
	7	69.9	3.56	68.9	2.07	17	68.9	2.07	24
	8	69.6	24.06	57.0	1.60	8	56.1	0.00	13
	9	76.3	1.73	75.9	1.20	18	75.5	0.67	14
	10	101.1	14.63	94.6	7.26	10	93.0	5.44	10
30	11	152.6	10.82	147.8	7.33	51	149.5	8.57	61
	12	86.4	6.14	83.3	2.33	17	82.5	1.35	36
	13	107.6	6.64	105.7	4.76	53	104.5	3.57	41
	14	113.2	16.94	105.8	9.30	22	113.2	16.94	41
	15	173.8	26.95	159.0	16.14	57	157.4	14.97	79
	16	127.2	19.77	118.5	11.58	40	119.5	12.52	51
	17	110.2	10.64	104.5	4.92	38	104.2	4.62	41
	18	131.4	11.54	125.5	6.54	20	131.2	11.38	46
	19	185.0	18.29	173.8	11.13	27	173.8	11.13	41
	20	140.5	11.86	135.2	7.64	58	138.3	10.11	93
40	21	261.3	57.69	215.0	29.75	311	226.7	36.81	209
	22	189.0	18.42	178.8	12.03	163	183.4	14.91	165
	23	325.7	76.05	273.9	48.05	315	264.3	42.86	373
	24	360.2	60.73	326.6	45.74	325	342.2	52.70	351
	25	162.0	21.53	155.1	16.35	206	154.8	16.13	140
	26	273.1	35.67	260.4	29.36	130	259.6	28.96	298
	27	233.0	35.31	200.8	16.61	209	215.8	25.32	282
	28	408.5	92.96	286.2	35.19	373	294.3	39.02	109
	29	268.4	48.86	219.4	21.69	202	223.4	23.90	175
	30	280.8	65.08	240.9	41.62	209	254.7	49.74	395
Avg.			23.99		13.32	98		14.69	105

instances with $n = 40$ vessels. The runtimes indicate that the two meta-heuristics are extremely useful for practice.

Summarizing, SWO and TS deliver solutions of near optimal quality for small-sized instances. While the average errors for medium-sized and large-sized instances still indicate a further optimization potential, the meta-heuristics are the only methods that deliver solutions of acceptable quality for these instances. The slightly better overall performance of SWO against TS in terms of solution quality as well as computation time makes it the preferable solution method for the BACAP. SWO is therefore used as the reference solution method in the following tests.

Test 6.4: Comparison with the Park–Kim approach

In order to further assess the quality of the approach, SWO is compared with the Lagrangean heuristics proposed by Park and Kim (2003). The authors report solutions and lower bounds for a set of 50 test instances with $n \in \{20, 25, 30, 35, 40\}$ vessels. Solving these instances by the new BACAP approach requires slight modifications. The indices for quay segments and for time periods are adapted and the objective function is replaced as defined in the model of Park and Kim (2003). To eliminate the influence of decreasing effects of QC productivity, the interference exponent is set to $\alpha = 1$ and the berth deviation factor is set to $\beta = 0$.

Table 6.6 shows the cost of the best solution found by Park and Kim for each of the test instances in column PK. It is compared with the solutions generated by the SWO heuristic. The table also shows the relative improvement *Impr* in percent as realized by SWO, i.e., $Impr = (\text{PK} - \text{SWO})/\text{PK} \times 100$. As can be seen, SWO always delivers better solutions for the 50 instances. On average, SWO improves

Table 6.6 Comparison of SWO with results of Park and Kim (2003)

$n = 20$				$n = 25$				$n = 30$			
#	PK	SWO	*Impr.*	#	PK	SWO	*Impr.*	#	PK	SWO	*Impr.*
1	53	42	20.75	11	85	80	5.88	21	109	98	10.09
2	93	87	6.45	12	126	113	10.32	22	221	194	12.22
3	161	145	9.94	13	145	135	6.90	23	190	166	12.63
4	91	77	15.38	14	64	58	9.38	24	77	71	7.79
5	78	74	5.13	15	86	73	15.12	25	174	161	7.47
6	31	27	12.90	16	163	147	9.82	26	130	117	10.00
7	93	75	19.35	17	127	118	7.09	27	103	90	12.62
8	47	41	12.77	18	142	134	5.63	28	171	144	15.79
9	65	52	20.00	19	69	60	13.04	29	230	188	18.26
10	156	145	7.05	20	213	199	6.57	30	94	78	17.02
Avg.			12.97				8.97				12.39

$n = 35$				$n = 40$			
#	PK	SWO	*Impr.*	#	PK	SWO	*Impr.*
1	158	136	13.92	11	181	162	10.50
2	138	123	10.87	12	219	200	8.68
3	136	124	8.82	13	313	239	23.64
4	208	181	12.98	14	234	222	5.13
5	245	203	17.14	15	333	301	9.61
6	169	150	11.24	16	269	238	11.52
7	187	167	10.70	17	271	240	11.44
8	196	175	10.71	18	215	188	12.56
9	172	151	12.21	19	250	217	13.20
10	197	168	14.72	20	359	274	23.68
Avg.			12.33				13.00

the objective function value by 12%. Park and Kim also report lower bounds for their test instances. Curiously, many of the solutions obtained by SWO fall below these bounds. Therefore the feasibility of every solution has been checked through a CPLEX analysis of Park and Kim's model by fixing the decision variables according to the SWO solution. CPLEX verifies that the found values of the decision variables are feasible with respect to the model. It also returns the same objective function value as SWO does. For this reason, either the lower bounds reported by Park and Kim are faulty, or their model implementation differs from their published mathematical formulation. Regardless of this open question, the gained results confirm the competitiveness of the new approach.

Test 6.5: Comparison with a sequential solution approach

In practice, BAP and QCAP are usually solved sequentially, whereas the BACAP provides an integrated solution. The following test compares these two alternatives. To simulate the sequential solution process, the procedure $QC\_Assignment(i, s_i, b_i)$ is removed from the procedure $Insert(i)$. Instead, the handling time of each vessel is fixed to the minimum handling time $d_i^{\min}$ as defined in (6.1) on page 57. Note that the terminal planners might apply handling times above $d_i^{\min}$ to anticipate terminal productivity influences. However, this anticipation is rather speculative and therefore not considered here. From the outlined modification, all presented solution procedures solve a BAP with fixed handling times. For the derived berth plan, a QCAP solution needs to be determined. To do so, the vessels are removed from the berth plan one by one and reinserted using the original procedure $Insert(i)$. The derived final solution comprises a berth plan and an assignment of QCs to vessels. The reinsertion ensures that the solution is feasible by revising berthing positions and berthing times whenever the QC assignment causes infeasibility of the original berth plan. Table 6.7 shows the derived objective function values if SWO is applied within the described sequential solution process in column SEQ. Column BACAP shows the objective function values if SWO is applied within the integrated solution process (these values are the same as reported in Table 6.5). Furthermore, the resulting relative improvement $Impr = (\text{SEQ} - \text{BACAP})/\text{SEQ} \times 100$ in percentage of the BACAP approach over the sequential solution approach is given.

As can be seen by the results, the integrated solution of the BACAP clearly dominates the sequential solution of BAP and QCAP. For small-sized and medium-sized instances the average improvement is about 12% and 16%, respectively. For the large-sized instances the average improvement is even larger than 40%. This shows that in a congested terminal the combined consideration of the affected resources quay space and QCs becomes an essential need. Their separate consideration within the BAP (quay space resource) and the QCAP (QC resource) leads to solutions of very poor quality.

The superior solution quality of the BACAP within all three instance sets indicates that deep integration of the BAP and the QCAP represents an advanced planning concept for seaside operations in a CT.

Table 6.7 Comparison with a sequential solution approach

$n = 20$				$n = 30$				$n = 40$			
#	SEQ	BACAP	*Impr.*	#	SEQ	BACAP	*Impr.*	#	SEQ	BACAP	*Impr.*
1	91.1	85.1	6.59	11	218.3	147.8	32.30	21	386.2	215.0	44.33
2	57.8	53.9	6.75	12	98.5	83.3	15.43	22	275.1	178.8	35.01
3	88.7	77.4	12.74	13	143.0	105.7	26.08	23	555.2	273.9	50.67
4	109.3	79.7	27.08	14	131.2	105.8	19.36	24	564.3	326.6	42.12
5	67.3	56.8	15.60	15	177.3	159.0	10.32	25	179.7	155.1	13.69
6	57.6	57.6	0.00	16	122.4	118.5	3.19	26	452.0	260.4	42.39
7	75.5	68.9	8.74	17	113.6	104.5	8.01	27	292.6	200.8	31.37
8	68.0	57.0	16.18	18	140.4	125.5	10.61	28	429.6	286.2	33.38
9	82.4	75.9	7.89	19	203.4	173.8	14.55	29	413.6	219.4	46.95
10	120.1	94.6	21.23	20	160.1	135.2	15.55	30	647.4	240.9	62.79
Avg.			12.28				15.54				40.27

Test 6.6: Effectiveness of vessel priorities

In the presented BACAP formulation, vessel priorities are modeled by different cost rates per vessel class. The solutions to the BACAP have to reflect these priorities to ensure satisfaction of vessel operators. To assess the effectiveness of the prioritization, the indicators of service quality, i.e., the speedup of vessels, the tardiness of vessels, and the deviations from desired berthing positions, are measured. If priority is given to certain vessels these values should decrease at the expense of vessels with lower priority.

For the test the benchmark suite is solved, once using identical cost rates for all vessel classes, i.e., no vessel receives priority, and once using the vessel class specific cost rates as stated in Table 6.2. The class specific cost rates give low priority to vessels belonging to the feeder class, medium priority to vessels belonging to the medium class, and high priority to vessels belonging to the jumbo class. Table 6.8 reports the average speedup (ΔETA_i), tardiness (ΔEFT_i), and berthing position deviation (Δb_i) of all vessels belonging to the same class as observed within the solutions to the 30 instances.

The results show that the specific cost rates for the vessel classes improve the service quality of jumbo vessels at the expense of feeder and medium vessels. For example, the average tardiness ΔEFT_i of a jumbo vessel decreases from 2.62 h in the solutions with identical cost rates to 1.13 h in the solutions with class specific cost rates. In contrast, the average tardiness of a feeder vessel increases from 0.34 to 0.64 h. Similar results can be observed regarding the speedup ΔETA_i of vessels. Interestingly, values observed for jumbo vessels are still larger than the corresponding values for feeder and medium vessels. An explanation is that speedups and tardiness of smaller vessels are frequently avoided by assigning them apart berthing positions. Due to the shorter vessel length, such alternative positions are often easier to find than for jumbo vessels. This is verified by the average deviation from desired

Table 6.8 Effects of vessel prioritization by cost rates

Cost rates	Identical			Class specific		
Vessel class	Feeder	Medium	Jumbo	Feeder	Medium	Jumbo
Avg. ΔETA_i	0.13	0.51	0.72	0.23	0.54	0.60
Avg. ΔEFT_i	0.34	1.06	2.62	0.64	1.11	1.13
Avg. Δb_i	5.83	4.70	3.68	6.01	4.76	2.23

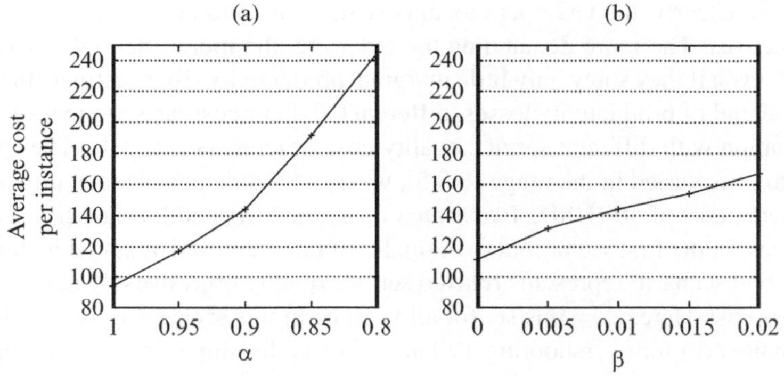

Fig. 6.9 Impact of α and β on average cost per instance

berthing positions which is 6.01 quay segments (≈60 m) for a feeder vessel but only 2.23 segments (≈22 m) for a jumbo vessel. The test shows that the incorporation of different cost rates for vessel classes is an effective way to prioritize vessels within the BACAP solution process.

Test 6.7: Estimating cost of productivity losses

In this test, the influence of QC productivity on the cost of solutions is investigated. To quantify its impact, α and β, previously set to 0.9 and 0.01, are varied separately. SWO is run for every parameter setting over all 30 instances. Figure 6.9a shows the average service cost obtained for the instances if α is varied from 1.0 to 0.8 and $\beta = 0.01$ is held constant. The inverse range is chosen to indicate that larger values of α correspond to smaller loss of crane productivity. The average cost per instance amounts to below 100,000 USD, if QC interference is neglected ($\alpha = 1$). With the still reasonable interference exponent $\alpha = 0.8$ it is more than doubled. Figure 6.9b shows the average cost obtained for the instances, if β is varied in the range from 0.00 to 0.02 and $\alpha = 0.9$ is held constant. Again, neglecting the impact of vessels' berthing positions on the crane productivity considerably underestimates costs. For $\beta = 0.02$, average cost are approximately 50% higher.

This result verifies the strong impact of crane productivity on the terminal cost. Incorporating or neglecting productivity losses in the berth planning is by no means a marginal difference. Providing reasonable productivity measures is therefore an important aspect of planning seaside operations in CTs.

Test 6.8: Effectiveness of QC operational cost consideration

Besides service quality cost of vessels also QC operational cost is included in the objective function of the BACAP. This cost reflects the CT management's desire to avoid productivity losses caused by crane interference and apart berthing positions of vessels. Clearly, the QC operational cost stays in conflict with the service quality objectives. The more demanding the latter are, the more often QCs need to be utilized, even if they show only little marginal productivity. To investigate the reduction potential of productivity losses, different QC-hour cost rates are investigated in combination with different service quality cost structures of vessels. The QC-hour cost rate c^4 is varied in the range [0,0.5], where $c^4 = 0.5$ means that a utilized QC-hour incurs cost of 500 USD. Two scenarios are investigated for the service quality objectives. In the first scenario the original cost rates are used as stated in Table 6.2. The second scenario represents relaxed service quality objectives by neglecting the tardiness cost. Here, $c_i^2 = 0$ is set for all vessels. In this scenario only speedup cost and penalty cost for overshooting the latest allowed finishing times LFT_i remain for the overall cost of service quality. Figure 6.10 shows the total observed productivity loss over all 30 instances for each combination of QC-hour cost rate and service quality scenario. The productivity loss is derived from the total utilized QC-hours in the solutions, minus the requested QC capacity (m_i) over all vessels contained within the instances.

As one would expect, the productivity loss decreases in both scenarios if the QC-hour cost rate c^4 is increased. The most considerable decrease is observed between

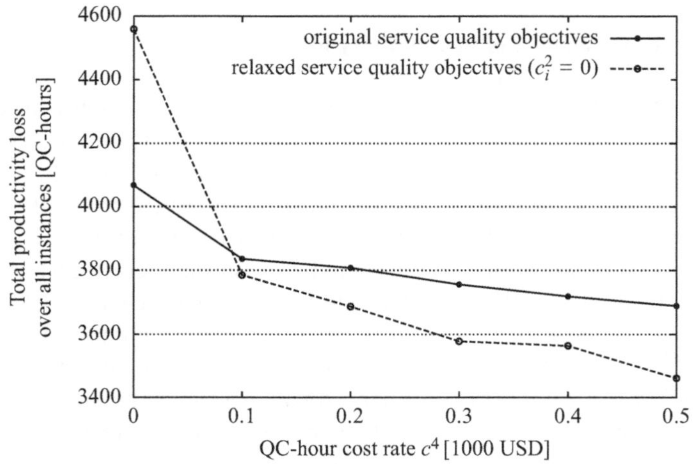

Fig. 6.10 Impact of QC operational cost on productivity loss

$c^4 = 0$ and $c^4 = 0.1$, which shows that even a small c^4 value avoids a waste of QC capacity. A further increase in c^4 leads to a further decrease in the productivity loss at a lower rate.

Focusing on the impact of service quality objectives, the relaxed scenario shows a higher productivity loss than the original scenario in the case of $c^4 = 0$. At first glance this is unexpected because the relaxation of service quality requirements allows to use QC capacity more effectively. However, with $c^4 = 0$ the solution methods do not aspire to minimize QC operational cost at all and thus, a reduction in productivity loss takes place only by chance. If c^4 takes a positive value, the productivity loss in the relaxed scenario is lower than in the scenario with original service quality objectives. This confirms that the service quality objectives and the QC operational cost objectives are in conflict and that the potential to reduce QC productivity loss depends on the service quality requirements.

Summarizing the test, incorporating QC operational cost in the BACAP provides an option to reduce crane productivity loss at a considerable rate. This result even holds in the presence of demanding service quality objectives.

Test 6.9: Potential of variable-in-time QC-to-Vessel assignments

The presented BACAP approach is able to deal with variable-in-time QC-to-Vessel assignments, i.e., the number of cranes serving a vessel may change during the service process. The studies of Oğuz et al. (2004), Liu et al. (2006), and Imai et al. (2008a) consider time-invariant QC-to-Vessel assignments only, i.e., the number of assigned cranes is held constant during the service of a vessel, which eases modeling and solving of seaside operations planning. A final test addresses the improvement potential offered by consideration of variable-in-time assignments. For comparison, SWO is restricted to consider time-invariant assignments only. This is realized by modifying the procedure *QC_Assignment* such that vessel i is served by a *constant* number of $q \in R_i$ cranes during its whole handling interval. This number is determined by the local refinement procedure for QC resource leveling.

Table 6.9 reports the objective function values obtained if SWO considers merely time-invariant QC-to-Vessel assignments in column INVAR. The results for the original SWO variant considering variable-in-time assignments are taken from Test 6.3 and appear in column VAR. The relative percentage improvement of VAR over INVAR is reported as $Impr = (\text{INVAR} - \text{VAR})/\text{INVAR} \times 100$.

The results show that consideration of variable-in-time QC-to-Vessel assignments leads to better solutions for all 30 instances. For small and medium-sized instances, the average improvements are within 6–7%. Interestingly, the instances with $n = 40$ vessels show an average improvement of 15%. This indicates that variable-in-time QC-to-Vessel assignments are of predominant importance in congested terminal situations. In such a situation a simultaneous service of vessels is made possible by removing and reassigning QCs in a flexible fashion. In contrast, under time-invariant QC-to-Vessel assignments, the service of a vessel must be postponed if only a subset of cranes is available.

Table 6.9 Results for time-invariant and for variable-in-time QC-to-Vessel assignments

$n = 20$				$n = 30$				$n = 40$			
#	INVAR	VAR	*Impr.*	#	INVAR	VAR	*Impr.*	#	INVAR	VAR	*Impr.*
1	89.0	85.1	4.38	11	148.0	147.8	0.14	21	293.2	215.0	26.67
2	56.2	53.9	4.09	12	93.4	83.3	10.81	22	193.8	178.8	7.74
3	89.8	77.4	13.81	13	111.2	105.7	4.95	23	331.4	273.9	17.35
4	81.8	79.7	2.57	14	115.8	105.8	8.64	24	366.0	326.6	10.77
5	59.2	56.8	4.05	15	175.8	159.0	9.56	25	171.6	155.1	9.62
6	59.2	57.6	2.70	16	126.6	118.5	6.40	26	278.4	260.4	6.47
7	75.8	68.9	9.10	17	114.6	104.5	8.81	27	235.8	200.8	14.84
8	61.4	57.0	7.17	18	144.4	125.5	13.09	28	412.4	286.2	30.60
9	79.0	75.9	3.92	19	180.8	173.8	3.87	29	255.0	219.4	13.96
10	105.0	94.6	9.90	20	139.8	135.2	3.29	30	284.2	240.9	15.24
Avg.			6.17				6.95				15.32

Summarizing, the improved solution quality offered by variable-in-time QC-to-Vessel assignments makes their consideration an essential need for seaside operations planning.

6.4 Summary

This chapter has provided a study on the combined Berth Allocation and Crane Assignment Problem (BACAP). The proposed mathematical formulation of the problem is able to tackle QC productivity losses caused by crane interference and caused by berthing vessels apart from desired berthing positions. It additionally comprises practical aspects such as the bounding of vessel speedups by earliest service start times and by taking care of QC operational cost in addition to common service quality objectives. Despite these extensions, the proposed mathematical formulation of the problem is more compact than the one presented in the pioneering work of Park and Kim (2003).

Several new heuristics have been presented and intensively tested. The computational results show that the local refinement procedures are effective in improving initial solutions. The two meta-heuristics SWO and TS both lead to further improvements. They deliver solutions of near optimal quality for small-sized instances and of reasonably good quality for medium-sized and large-sized instances. The computation times are acceptable even for large-sized instances. The tests also confirm the superiority of SWO over the solution method proposed in Park and Kim (2003) and the superiority of the integrated solution of BAP and QCAP over a sequential solution process. Further tests have revealed the strong impact of QC productivity effects on the obtained solutions, the effectiveness of the combined service quality and QC operational cost objectives, and the high quality solution potential offered by considering variable-in-time QC-to-Vessel assignments.

From these results, it is concluded that the influence of crane assignment and crane productivity is not marginal and needs to be considered as an essential input for berth planning. The parameters used to model the productivity losses and the cost rates used to model vessel priorities as well as QC operational cost objectives allow the CT management to adapt the BACAP flexibly on terminal specific characteristics such as the workload situation. The deep integration of BAP and QCAP proves to be a successful first step towards an integrated planning of seaside operations.

Chapter 7
Quay Crane Scheduling

This chapter deals with the QCSP on the basis of container groups. It is studied as an isolated problem here and functionally integrated into the BACAP in the next chapter. Crane scheduling for container groups has been introduced by Kim and Park (2004). As noted by Moccia et al. (2006), the original QCSP model provided by Kim and Park shows an inaccuracy regarding the detection of crane interference. Unfortunately, even reworked problem formulations still tolerate certain cases of crane interference. A corrected problem formulation and a heuristic solution method have been provided by Bierwirth and Meisel (2009). The model and the heuristic are presented in Sects. 7.1 and 7.2, respectively. In Sect. 7.3 the QCSP is extended by incorporation of time windows for the cranes. Necessary modifications of the mathematical formulation and the solution method are described. Computational tests follow in Sect. 7.4. The study on the QCSP is concluded in Sect. 7.5.

7.1 Modeling the QCSP

7.1.1 Problem Description and Assumptions

In the QCSP for container groups a set of tasks $\Omega = \{1,2,\ldots,n\}$ and a set of QCs $Q = \{1,2,\ldots,q\}$ are given. Each task $i \in \Omega$ represents a loading or unloading operation of a certain container group. The tasks have individual processing times p_i and bay positions l_i. Additionally, dummy tasks 0 and $T = n+1$ with processing times $p_0 = p_T = 0$ are given to indicate the begin and the end of the service of the vessel. Further task sets are defined by $\Omega^0 = \Omega \cup \{0\}$, $\Omega^T = \Omega \cup \{T\}$, and $\overline{\Omega} = \Omega \cup \{0,T\}$. Precedence relations may exist between pairs of tasks that are located within the same bay. Let Φ denote the set of precedence constrained task pairs. Furthermore, let $\Psi \supseteq \Phi$ denote the set of all task pairs for which it is known in advance that

F. Meisel, *Seaside Operations Planning in Container Terminals*,
Contributions to Management Science,
DOI: 10.1007/978-3-7908-2191-8_7, © Physica-Verlag Berlin Heidelberg 2009

they cannot be processed simultaneously. For each crane $k \in Q$ a ready time r^k and an initial bay position l_0^k is given. Without loss of generality, it is assumed that the cranes are indexed sequentially according to their initial positioning alongside the vessel. All QCs can move between two adjacent bays in an identical travel time $\hat{t} > 0$. It is supposed that no two QCs can operate at the same bay at the same time. Moreover, cranes are not allowed to cross each other and have to keep a safety margin δ, measured in units of bays. The problem is to find completion times c_i for all tasks $i \in \overline{\Omega}$ on the cranes with respect to the constraints, such that the completion time c_T of the final task T (i.e., the makespan) is minimized.

Assumptions of the QCSP with container groups are:

1. Container groups are predefined from given stowage plans.
2. Processing of tasks is non-preemptive.
3. All QCs show an identical, deterministic transshipment productivity. For this reason fixed processing times of tasks are given. No consideration of individual container moves or crane cycle times is necessary.
4. The order of processing the tasks of a bay is completely determined by precedence constraints.
5. Crane operations cannot lead to an instable load configuration of the vessel, i.e., stability issues are not considered in the QCSP.
6. There is sufficient space beside the vessel to place idle QCs outside of the vessel area.

The described problem corresponds to a minimum makespan scheduling problem with parallel identical machines and precedence constraints. This problem is known to be $\mathcal{NP}$-hard in the strong sense, provided that more than two machines, non-preemption or non-uniform processing times are given, see Pinedo (2002).

7.1.2 Conventional Formulation of Interference Constraints

In the BACAP study, the productivity loss caused by crane interference has been modeled using an interference exponent α. Within the QCSP, interference effects are considered in more detail in order to generate feasible QC schedules.

In correspondence with models in machine scheduling, it is supposed that no two QCs can operate at the same bay at the same time. Moreover, since QCs are rail mounted, two types of interference constraints have to be respected:

- Non-crossing constraint: QCs cannot cross each other.
- Safety constraint: Adjacent QCs have to keep a safety margin at all times.

The *safety margin* δ signifies a certain number of in-between bays between adjacent QCs. If, for example, $\delta = 1$ this means that QCs can simultaneously operate at the vessel if they are separated by at least one bay. Safety margin must be respected not only while QCs are working but also during the movement operations.

Interference constraints were first included in the QCSP model of Kim and Park (2004). As noted by Moccia et al. (2006) this model does not detect interference in every case. Therefore, Moccia et al. (2006) have revised the model of Kim and Park by incorporating travel times for QCs that subsequently process tasks in the same bay. A compact mathematical formulation of the revised model can be found in Sammarra et al. (2007). Unfortunately, also the modification proposed in the revised model may yield solutions where QCs cross or violate the safety margin. To demonstrate the incorrectness of this model those constraints that are responsible for detecting crane interference, labeled (9)–(14) in the paper of Sammarra et al. (2007), are briefly revisited:

$$c_i + p_j - c_j \leq M(1 - z_{ij}) \qquad \forall i, j \in \Omega, \tag{7.1}$$

$$c_i + p_j - c_j + \sum_{k \in Q} \sum_{\substack{u \in \Omega^0, \\ l_u \neq l_i}} \hat{t} x^k_{uj} \leq M(1 - z_{ij}) \qquad \forall i, j \in \Omega, l_i = l_j, \tag{7.2}$$

$$c_j - p_j - c_i \leq M z_{ij} \qquad \forall i, j \in \Omega, \tag{7.3}$$

$$c_j - p_j - c_i - \sum_{k \in Q} \sum_{\substack{u \in \Omega^0, \\ l_u \neq l_i}} \hat{t} x^k_{uj} \leq M z_{ij} \qquad \forall i, j \in \Omega, l_i = l_j, \tag{7.4}$$

$$z_{ij} + z_{ji} = 1 \qquad \forall (i, j) \in \Psi, \tag{7.5}$$

$$\sum_{v=1}^{k} \sum_{u \in \Omega^0} x^v_{uj} - \sum_{v=1}^{k} \sum_{u \in \Omega^T} x^v_{ui} \leq M(z_{ij} + z_{ji}) \quad \forall i, j \in \Omega, l_i < l_j, \forall k \in Q. \tag{7.6}$$

In this formulation x^k_{ij} and z_{ij} denote binary decision variables. x^k_{ij} is set to 1 if tasks i and j are processed consecutively by crane k, and z_{ij} is set to 1 if task j starts after the completion of task i. The variables z_{ij} are defined in Constraints (7.1) and (7.3). Constraints (7.2) and (7.4) ensure that the travel time $\hat{t}$ is kept between the completion of a task and the start of the next task in the same bay if both tasks are processed by different QCs. In order to express a safety margin of one bay between adjacent QCs, Sammarra et al. (2007) include those pairs of tasks in set Ψ that belong to adjacent bays. Constraints (7.5) ensure that these tasks are not processed simultaneously. Finally, a simultaneous processing of tasks that inevitably requires a crossing of the assigned cranes is forbidden by Constraints (7.6).

To demonstrate the incorrectness of the above interference constraints two small example problems are considered which are solved infeasible. The first problem consists of four tasks with processing times 10, 20, 40, and 30, positioned in bays 1, 3, 5, and 7, respectively, and two cranes. The safety margin is set to one bay and the travel time to $\hat{t} = 1$ time unit per bay. Figure 7.1a shows an optimal schedule derived from the above model. In this solution, tasks 1 and 3 are assigned to QC 1, and tasks 2 and 4 are assigned to QC 2. Obviously, the solution is infeasible because QC 1 crosses QC 2 in order to process task 3. Since no two tasks are positioned within the same bay or adjacent bays, $\Psi = \emptyset$ in this problem. Therefore, Constraints (7.2), (7.4), and (7.5) do not appear in the model instance. The start

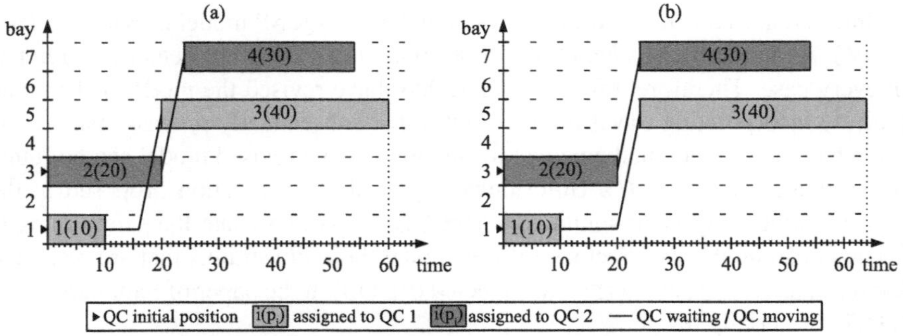

Fig. 7.1 Violation of the non-crossing requirement (**a**) and a feasible schedule (**b**)

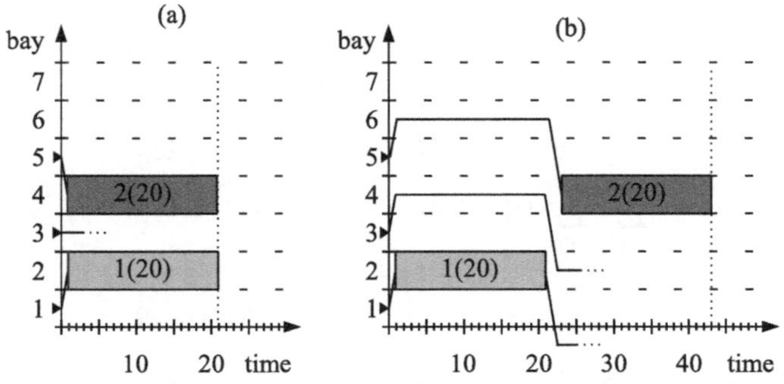

Fig. 7.2 Violation of the safety margin (**a**) and a feasible schedule (**b**)

time of task 4 is derived from the completion time of task 2 plus the time needed by QC 2 to travel from bay 3 to bay 7. Furthermore, the start time of task 3 is delayed by Constraint (7.1) because tasks 2 and 3 are not allowed to be processed simultaneously, as effected by setting $z_{23} = 1$ in Constraint (7.6). However, the inserted delay is insufficient to avoid a crossing of the cranes. The corrected feasible solution is shown in Fig. 7.1b. This solution is obtained by introducing a temporal distance of four time units between the completion of task 2 and the start of task 3. Through this correction the makespan increases from 60 to 64 time units.

The model formulation encounters a further weakness regarding QCs that stay idle during the service. The second problem instance consists of two tasks with identical processing time, positioned in bays 2 and 4. Three QCs, initially positioned at bays 1, 3, and 5, are available for the service. The safety margin and the travel time of cranes are as above. Figure 7.2a shows one optimal schedule where the QC positioned initially at bay 1 processes task 1 and the QC positioned at bay 5 processes task 2. Since there are no restrictions on idle cranes in the above model, the shown solution is not forbidden. However, it is infeasible because the idle QC is within the safety area of the active QCs. Without changing the task-to-QC assignment, a corrected feasible solution requires processing of the tasks consecutively. Moreover, as shown in Fig. 7.2b, the starting time of task 2 needs a further delay of two time units to enable a safe movement of the cranes. Now the safety margin is always kept during the entire operation. As in the first problem, the inclusion of a suitable temporal distance between tasks resolves the crane conflict. Unfortunately, the repair does not preserve optimality of the solution.

7.1.3 Corrected Formulation of Interference Constraints

The above analysis discloses a serious weakness in the existing QCSP models. Temporal distances between tasks are only included if these tasks are positioned within the same bay. The key to a correct model formulation is the determination of a suitable temporal distance between any two tasks involved in a problem. For this purpose, the temporal distance is computed as a function of the bay positions of tasks, the safety margin, the QC travel time, and, last but not least, the realized task-to-QC assignment.

Let Δ_{ij}^{vw} denote the minimum time span to elapse between the processing of two tasks i and j, if processed by QCs v and w respectively. Due to the sequential indexing of cranes, one can say that v operates below (above) w if $v < w$ ($v > w$). Generally, a crossing must be avoided if the lower QC processes a task which is located at a bay above a task processed by the upper QC. Furthermore, compliance of the safety margin must be guaranteed between any two cranes v and w. Let δ_{vw} be the smallest allowed difference between the bay positions of cranes v and w. It is calculated as

$$\delta_{vw} = (\delta + 1)|v - w|. \tag{7.7}$$

For all combinations of tasks $i, j \in \Omega$ and QCs $v, w \in Q$ the minimum temporal distance is now defined as

$$\Delta_{ij}^{vw} = \begin{cases} (l_i - l_j + \delta_{vw})\hat{t}, & \text{if } v < w \text{ and } i \neq j \text{ and } l_i > l_j - \delta_{vw}, \\ (l_j - l_i + \delta_{vw})\hat{t}, & \text{if } v > w \text{ and } i \neq j \text{ and } l_i < l_j + \delta_{vw}, \\ 0, & \text{otherwise.} \end{cases} \tag{7.8}$$

The first case of (7.8), in which v operates below w, is illustrated in Fig. 7.3. Here, tasks i and j are processed by adjacent QCs v and $w = v + 1$. In the example, the safety margin is set to $\delta = 2$ and the QC travel time is set to $\hat{t} = 1$. Assuming that task j is completed at time c_j, QC v must not be positioned above $l_j - \delta_{vw}$ at this point in time because it operates below w. Since v processes its next task at bay l_i it has to traverse at least $l_i - (l_j - \delta_{vw})$ bays. The resulting minimum travel time of v is $\Delta_{ij}^{vw} = (l_i - l_j + \delta_{vw})\hat{t} = 6$ with $l_i - l_j = 3$. Note, that Δ_{ij}^{vw} yields the same value if task i is processed prior to j under ceteris paribus conditions. The reverse positioning of QCs is treated in the second case of (7.8). Without loss of generality, every instance of this case can be transformed into an identical instance of the former case by exchanging the roles of tasks i and j. In all other cases, cranes cannot come into conflict as indicated by setting the temporal distance to a value of 0.

The example in Fig. 7.3 considers interference of adjacent QCs. However, if v and w are not adjacent, (7.8) calculates a sufficiently large temporal distance between the

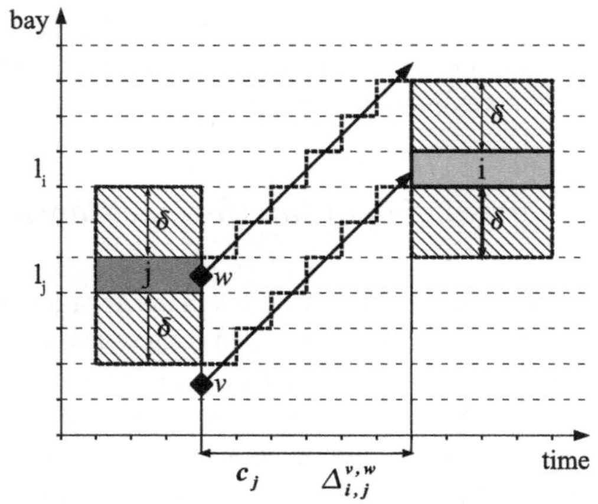

Fig. 7.3 Necessary time span between the execution of tasks by adjacent QCs

processing of any two tasks permitting a safe movement of the in-between cranes. This also applies if in-between cranes are idle, see Fig. 7.2b.

Let Θ denote the set of all combinations of tasks and QCs that potentially lead to crane interference. It can be defined as

$$\Theta = \{(i, j, v, w) \in \Omega^2 \times Q^2 \mid (i < j) \wedge (\Delta_{ij}^{vw} > 0)\}. \tag{7.9}$$

Due to the symmetry of the temporal distances Δ_{ij}^{vw} the consideration can be restricted to pairs of tasks with $i < j$. Actually, a certain combination in Θ will cause interference only if its task-to-QC assignment is selected in the QC schedule. Since this is unknown in advance, every element of Θ must be treated by a constraint in the QCSP model.

A correct formulation of the model of Sammarra et al. (2007) results, if Constraints (7.2), (7.4), and (7.6) are replaced by the following constraints:

$$\sum_{u \in \Omega^0} x_{ui}^v + \sum_{u \in \Omega^0} x_{uj}^w \leq 1 + z_{ij} + z_{ji} \qquad \forall (i, j, v, w) \in \Theta, \tag{7.10}$$

$$c_i + \Delta_{ij}^{vw} + p_j - c_j \leq M\Big(3 - z_{ij} - \sum_{u \in \Omega^0} x_{ui}^v - \sum_{u \in \Omega^0} x_{uj}^w\Big) \qquad \forall (i, j, v, w) \in \Theta, \tag{7.11}$$

$$c_j + \Delta_{ij}^{vw} + p_i - c_i \leq M\Big(3 - z_{ji} - \sum_{u \in \Omega^0} x_{ui}^v - \sum_{u \in \Omega^0} x_{uj}^w\Big) \qquad \forall (i, j, v, w) \in \Theta. \tag{7.12}$$

In Constraints (7.10) those assignments of tasks to QCs are identified that are realized in the schedule. Here, $\sum_{u \in \Omega^0} x_{ui}^v = 1$ if and only if task i is processed by QC v and $\sum_{u \in \Omega^0} x_{uj}^w = 1$ if and only if task j is processed by QC w. If both assignments take place, the left side reveals a value of two and the tasks are not allowed to be processed simultaneously, i.e., either $z_{ij} = 1$ or $z_{ji} = 1$. In the case of $z_{ij} = 1$ Constraints (7.11) insert the minimum temporal distance calculated by (7.8) between the completion time of task i and the starting time of task j. The corresponding case of $z_{ji} = 1$ is handled in Constraints (7.12).

7.1.4 Optimization Model

In the original model of Kim and Park (2004) the minimization of the weighted sum of makespan and QC finishing times is pursued. Due to the predominant importance of short vessel handling times in current CT markets, most authors merely consider the minimization of makespan, as has been done in the computational studies of Kim and Park (2004), Moccia et al. (2006), and Sammarra et al. (2007) too. The following formulation takes up this lead and ignores QC finishing times in the objective function.

To model the movement of cranes, the travel time of QC k to traverse from its initial position l_0^k to l_j ($j \in \Omega$) is defined as $t_{0j}^k = \hat{t}|l_0^k - l_j|$. The travel time between bay positions l_i and l_j ($i, j \in \Omega$) is defined as $t_{ij} = \hat{t}|l_i - l_j|$. If i or j or both belong to $\{0, T\}$ the travel time is set to $t_{ij} = 0$. Since the minimization of QC finishing times is ignored here, final repositioning movements of QCs after completion of the vessel's service are not considered in this formulation.

The QCSP is formulated as follows:

$$\text{minimize } c_T \tag{7.13}$$

subject to

$$\sum_{j \in \Omega^T} x_{0j}^k = 1 \qquad \forall k \in Q, \tag{7.14}$$

$$\sum_{j \in \Omega^0} x_{jT}^k = 1 \qquad \forall k \in Q, \tag{7.15}$$

$$\sum_{k \in Q} \sum_{j \in \Omega^T} x_{ij}^k = 1 \qquad \forall i \in \Omega, \tag{7.16}$$

$$\sum_{j \in \Omega^0} x_{ji}^k - \sum_{j \in \Omega^T} x_{ij}^k = 0 \qquad \forall i \in \Omega, \forall k \in Q, \tag{7.17}$$

$$c_i + t_{ij} + p_j - c_j \leq M(1 - x_{ij}^k) \qquad \forall i, j \in \overline{\Omega}, \forall k \in Q, \tag{7.18}$$

$$c_i + p_j - c_j \leq 0 \qquad \forall (i, j) \in \Phi, \tag{7.19}$$

$$c_i + p_j - c_j \leq M(1 - z_{ij}) \qquad \forall i, j \in \Omega, \tag{7.20}$$

$$c_j - p_j - c_i \leq M z_{ij} \qquad \forall i, j \in \Omega, \tag{7.21}$$

$$z_{ij} + z_{ji} = 1 \qquad \forall (i, j) \in \Psi, \tag{7.22}$$

$$\sum_{u \in \Omega^0} x_{ui}^v + \sum_{u \in \Omega^0} x_{uj}^w \leq 1 + z_{ij} + z_{ji} \qquad \forall (i, j, v, w) \in \Theta, \tag{7.23}$$

$$c_i + \Delta_{ij}^{vw} + p_j - c_j \leq M(3 - z_{ij} - \sum_{u \in \Omega^0} x_{ui}^v - \sum_{u \in \Omega^0} x_{uj}^w) \qquad \forall (i, j, v, w) \in \Theta, \tag{7.24}$$

$$c_j + \Delta_{ij}^{vw} + p_i - c_i \leq M(3 - z_{ji} - \sum_{u \in \Omega^0} x_{ui}^v - \sum_{u \in \Omega^0} x_{uj}^w) \qquad \forall (i, j, v, w) \in \Theta, \tag{7.25}$$

$$r^k + t^k_{0j} + p_j - c_j \leq M(1 - x^k_{0j}) \qquad \forall j \in \Omega, \forall k \in Q, \qquad (7.26)$$
$$c_i \geq 0 \qquad \forall i \in \overline{\Omega}, \qquad (7.27)$$
$$x^k_{ij} \in \{0,1\} \qquad \forall i,j \in \overline{\Omega}, \forall k \in Q, \qquad (7.28)$$
$$z_{ij} \in \{0,1\} \qquad \forall i,j \in \Omega. \qquad (7.29)$$

The pursued objective given in (7.13) is to minimize the handling time of the vessel referred to as the makespan. The makespan is defined by the completion time of dummy task T because every crane is enforced to visit this task after processing its assigned non-dummy tasks. Constraints (7.14) and (7.15) ensure that every QC starts with the initial dummy task 0 and ends up with the final dummy task T. If both fall together, i.e., $x^k_{0T} = 1$, QC k remains idle during the entire service. Constraints (7.16) ensure that each non-dummy task is processed exactly once. Constraints (7.17) ensure that every non-dummy task has a preceding task and a succeeding task. The completion times of the tasks are computed in Constraints (7.18) where M is again a sufficiently large positive number. Note that for $j = T$ the makespan is computed by this constraint. The precedence relations are included in Constraints (7.19) with respect to the task completion times. Constraints (7.20) and (7.21) set the variables z_{ij}. On this basis the non-simultaneity condition of tasks is represented in Constraints (7.22). Constraints (7.23)–(7.25) are the new interference constraints formulated in the previous section. The ready times of QCs are handled in (7.26) and the feasible domains of the decision variables are defined in (7.27)–(7.29).

The number of variables used in this formulation grows in $O(n^2q)$, which is the same as in Sammarra et al. (2007). Due to the newly formulated interference handling, the number of constraints grows in $O(n^2q^2)$ instead of $O(n^2q)$.

The QCSP formulation is classified using the scheme of Sect. 4.2.1. Tasks are defined by container groups (*Group*) where precedence relations exist among pairs of tasks (*prec*). The model considers ready times, initial positions, and movement time of cranes, classified by the crane attribute values *ready*, *pos*, and *move*. The non-crossing requirement and safety margins are respected (*cross*, *save*). The makespan, i.e., the maximum completion time (*compl*) among tasks, is minimized. Hence, the model is classified by $Group, prec \mid ready, pos, move \mid cross, save$ $max(compl)$.

Example 7.1: QCSP instance: definition and optimal solution

Table 7.1 shows the data of a small QCSP instance, which is used in the following to illustrate the proposed solution procedure. The problem contains nine container groups placed in a vessel with eleven bays. Two QCs are assigned to this vessel. The minimum temporal distances for combinations of task pairs and QC pairs are preprocessed according to (7.8). E.g., for $(i,j,v,w) = (7,8,2,1)$ one obtains $\delta_{21} = 2$ and $\Delta^{21}_{78} = (9-7+2) = 4$. The complete matrix Δ^{vw}_{ij} is shown in Table 7.2. From this it can be seen that Δ is symmetric in i,j and v,w, i.e., $\Delta^{vw}_{ij} = \Delta^{wv}_{ji}$. Set Θ is composed

Table 7.1 Example QCSP instance

Task index i	1	2	3	4	5	6	7	8	9
Processing time p_i	22	46	8	70	10	38	40	16	12
Bay position l_i	1	1	2	3	5	5	7	9	11
Precedence-constrained tasks	$\Phi = \{(1,2),(5,6)\}$								
Non-simultaneous tasks	$\Psi = \{(1,2),(1,3),(2,3),(3,4),(5,6)\}$								
QC 1	$l_0^1 = 1, \quad r^1 = 0$								
QC 2	$l_0^2 = 4, \quad r^2 = 0$								
QC travel speed	$\hat{t} = 1$								
Safety margin	$\delta = 1$								

Table 7.2 Obtained temporal distances Δ_{ij}^{vw} for the QCSP instance

j	w \\ i / v	1 1	 2	2 1	 2	3 1	 2	4 1	 2	5 1	 2	6 1	 2	7 1	 2	8 1	 2	9 1	 2
1	1	0	0	0	2	0	1	0	0	0	0	0	0	0	0	0	0	0	0
	2	0	0	2	0	3	0	4	0	6	0	6	0	8	0	10	0	12	0
2	1	0	2	0	0	0	1	0	0	0	0	0	0	0	0	0	0	0	0
	2	2	0	0	0	3	0	4	0	6	0	6	0	8	0	10	0	12	0
3	1	0	3	0	3	0	0	0	1	0	0	0	0	0	0	0	0	0	0
	2	1	0	1	0	0	0	3	0	5	0	5	0	7	0	9	0	11	0
4	1	0	4	0	4	0	3	0	0	0	0	0	0	0	0	0	0	0	0
	2	0	0	0	0	1	0	0	0	4	0	4	0	6	0	8	0	10	0
5	1	0	6	0	6	0	5	0	4	0	0	0	2	0	0	0	0	0	0
	2	0	0	0	0	0	0	0	0	0	0	2	0	4	0	6	0	8	0
6	1	0	6	0	6	0	5	0	4	0	2	0	0	0	0	0	0	0	0
	2	0	0	0	0	0	0	0	0	2	0	0	0	4	0	6	0	8	0
7	1	0	8	0	8	0	7	0	6	0	4	0	4	0	0	0	0	0	0
	2	0	0	0	0	0	0	0	0	0	0	0	0	0	0	4	0	6	0
8	1	0	10	0	10	0	9	0	8	0	6	0	6	0	4	0	0	0	0
	2	0	0	0	0	0	0	0	0	0	0	0	0	0	0	0	0	4	0
9	1	0	12	0	12	0	11	0	10	0	8	0	8	0	6	0	4	0	0
	2	0	0	0	0	0	0	0	0	0	0	0	0	0	0	0	0	0	0

according to Definition (7.9). In the example it consists of 41 elements indicated by the positive entries below the main diagonal of matrix Δ.

The optimal solution to the problem is given by the task sequences 0-1-2-3-6-8-10 for QC 1 and 0-4-5-7-9-10 for QC 2. The associated task completion times are $c_i = (0,\ 22,\ 68,\ 80,\ 71,\ 83,\ 123,\ 125,\ 145,\ 141,\ 145)$. The corresponding schedule is depicted in Fig. 7.4. It respects the required temporal distances between task pairs $(3,4)$, $(5,6)$, and $(7,8)$ in order to avoid violations of constraints. Note that QC 1 becomes idle three times because it has to wait until QC 2 has finished a task and moves away.

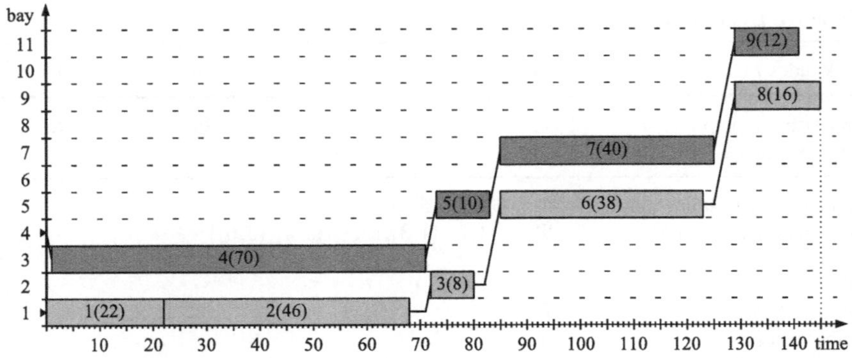

Fig. 7.4 Optimal QC schedule with makespan $c_T = 145$

7.2 Unidirectional Scheduling Heuristic

7.2.1 Idea and Outline

Following Bierwirth and Meisel (2009) a QC schedule is called a *unidirectional schedule* if the QCs do not change in moving direction after the initial repositioning and have identical directions of movement either from upper to lower bays or vice versa. The schedule depicted in Fig. 7.4 is a unidirectional schedule. As shown by Lim et al. (2007) for the QCSP with tasks defined by complete bays, there is at least one optimal schedule among the unidirectional ones. For this reason, anchoring unidirectionality in mathematical models of this QCSP variant, as has been done by Liu et al. (2006) and Lim et al. (2007), does not exclude the optimal schedule from the solution space.

In contrast, for the QCSP with container groups, one can easily construct instances with no unidirectional schedule existing among the optimal solutions. One such example is shown in Fig. 7.5 where (a) shows the optimal schedule and (b) and (c) show best unidirectional schedules. Note that the optimal schedule cannot be transformed into a unidirectional schedule without increasing the makespan or violating the precedence constraints of tasks in bay 3. Therefore, an optimization model for the QCSP with container groups must not restrict the structure to unidirectional schedules.

Although searching the space of unidirectional schedules can fail in finding the optimal solution, it might be a good strategy for solving the QCSP with container groups heuristically. The basic idea of the proposed heuristic is to search the space of unidirectional schedules exhaustively, ending up with an optimal schedule among the unidirectional ones. The *Unidirectional Scheduling (UDS) heuristic* respects all requirements of the QCSP including the issue of crane interference. It generates schedules by making decisions at three distinct levels:

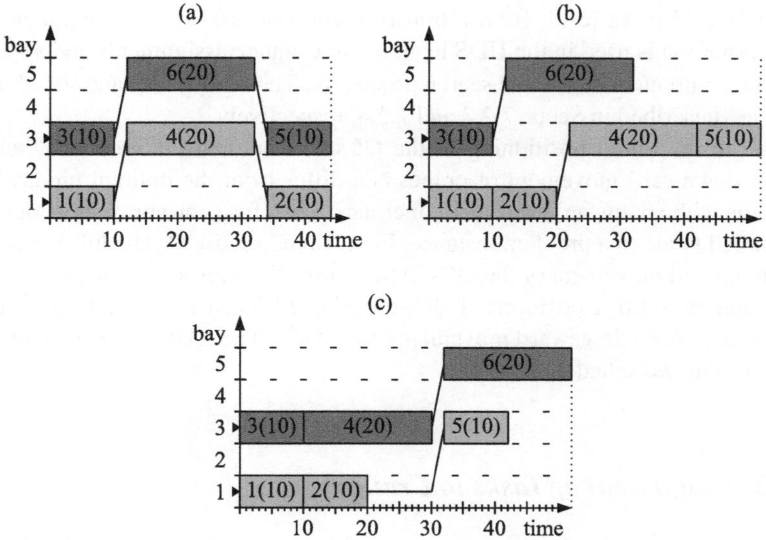

Fig. 7.5 A QCSP instance with a non-unidirectional optimal schedule (**a**) and best unidirectional schedules (**b**, **c**)

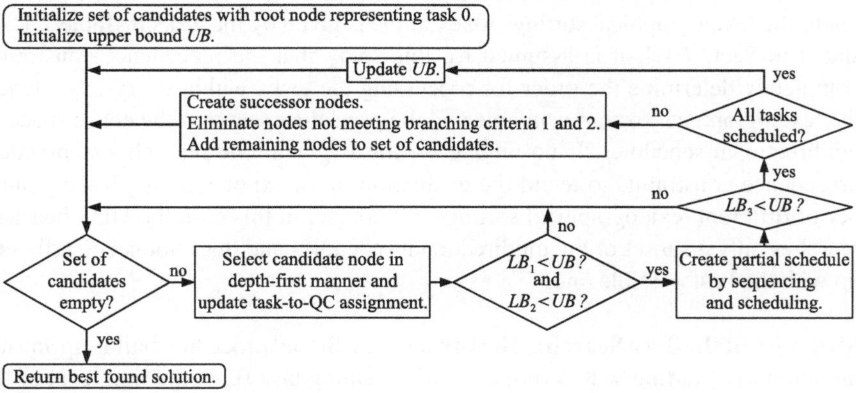

Fig. 7.6 Flowchart of the tree search

1. Task-to-QC Assignment: Using a tree search, the possible assignments of tasks to QCs are generated that allow for a unidirectional schedule.
2. Sequencing of tasks: For every considered task-to-QC assignment, sequences of tasks are determined that can be processed by QCs in a unidirectional movement.
3. Schedule building: Starting times for the tasks are iteratively determined with respect to the task sequences and the required temporal distances.

Task-to-QC assignments are the most complex decisions that need to be made if a unidirectional schedule is searched, see Lim et al. (2007). The UDS heuristic applies a tree search to generate these assignments as shown in Fig. 7.6. A detailed description of the procedure is given in Sect. 7.2.2. Decisions regarding the sequencing and scheduling of tasks result from a transformation of task-to-QC assignments. The transformation is used in the UDS heuristic to evaluate assignments and appears as a single component in the tree search procedure. The sequencing and scheduling of tasks are described in Sects. 7.2.3 and 7.2.4, respectively.

Due to the initial positioning of the QCs, the optimal unidirectional schedule with a downward movement of cranes can differ from the optimal unidirectional schedule with an upward movement of cranes. Therefore, the above procedure must be applied twice to a problem instance. First, a unidirectional schedule is generated for an upward movement of the QCs. Afterwards, the bays are numbered in inverse order and the starting positions of QCs are adapted. Then a unidirectional schedule is generated for a downward movement of the QCs. The UDS heuristic delivers the better of the two schedules.

7.2.2 Assignment of Tasks to Cranes

Lexicographical Sorting: In the first step of the assignment procedure, the tasks involved in a problem are sorted in lexicographical order of increasing bay position and, within a bay, by precedence relations. Afterwards, the tasks are indexed according to the lexicographical sorting. An example is given by the tasks in Table 7.1. As stated in Sect. 7.1.1, it is assumed for this study that the precedence constraints completely determine the order for processing the tasks within every bay. Hence, the lexicographical sorting is unique and the procedure searches the entire space of unidirectional schedules. If no unique sorting is given, one may choose practical precedence constraints to avoid the evaluation of an exponentially growing number of different lexicographical sortings. Of course, in this case the UDS heuristic searches only a subset of the unidirectional schedules and does not necessarily end up with the best possible one.

Structure of the Tree Search: The Branch-and-Bound procedure builds up an enumeration tree starting with a root node representing task 0. At the first level of the tree, task 1 is assigned to QC $m_1 \in Q$. At tree level 2, task 2 is assigned to QC $m_2 \in Q$, and so on. Figure 7.7 shows a sketch of the search tree for the QCSP instance in Table 7.1. Since nine tasks are assigned to two QCs, each path from the root to a leaf in the tree corresponds to one of the 2^9 task-to-QC assignments. The bold printed path in the tree represents the task-to-QC assignment that underlies the schedule depicted in Fig. 7.4.

To calculate an initial upper bound UB for the tree search, two task-to-QC assignments are generated using the *S-TASKS* rule and the *S-LOAD* rule, as introduced by Sammarra et al. (2007). They divide the entire task set into q subsets such that the number of tasks (*S-TASKS*) or the workload (*S-LOAD*) is almost equally shared among the cranes. The delivered task-to-QC assignments are evaluated by the sequencing and scheduling procedures as described later in Sects. 7.2.3 and

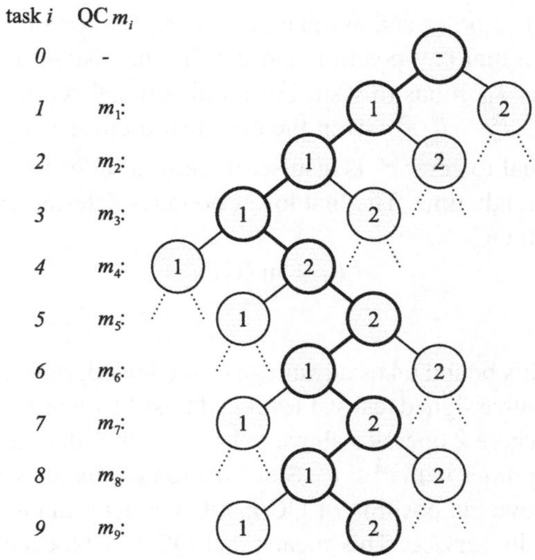

Fig. 7.7 Sketch of the search tree indicating the task-to-QC assignment

7.2.4. The shorter makespan computed for the two assignments serves as the initial upper bound.

The tree is searched in depth-first manner. A node is inspected by computing up to three lower bounds for the partial assignment. It is pruned if a lower bound overshoots UB. Otherwise, branching criteria are applied to decide on the successor nodes of the current node. Nodes that have passed the criteria enter the set of candidates that require further inspection. The upper bound UB is updated if a new best schedule has been generated. The Branch-and-Bound procedure terminates if the set of candidates is empty.

Bounding: The unique path leading from the root to a certain node in the search tree represents a partial task-to-QC assignment. Three lower bounds are applied to decide whether the current node at level i is pruned or not.

Lower bound 1: A lower bound on the makespan for the addressed partial task-to-QC assignment is computed by estimating the points in time when the cranes finish service. The finishing time c^k of QC $k \in Q$ is estimated by

$$c^k = r^k + w^k + d^k. \tag{7.30}$$

Here, r^k denotes the ready time of QC k, w^k denotes the workload of QC k under the current partial task-to-QC assignment, and d^k denotes the minimum travel time of QC k to traverse between the bays. The workload of QC k is computed by $w^k = \sum_{j \in \Omega^k} p_j$, where Ω^k denotes the set of all tasks prior to or equal to task i

in the lexicographical order and assigned to QC k. The travel time of the crane is determined by its initial bay position and the distance between the lowest bay l_{lo} and the upmost bay l_{up} it has to visit. For unidirectional schedules it is calculated as $d^k = (|l_0^k - l_{\mathrm{lo}}^k| + l_{\mathrm{up}}^k - l_{\mathrm{lo}}^k)\hat{t}$. Given the case that a crane is completely idle, i.e., its workload is equal to zero, c^k is also set to zero in order to neglect the possible influence of a late ready time. The first lower bound is determined by the maximum finishing time of all QCs

$$LB_1 = \max_{k \in Q}\{c^k\}. \tag{7.31}$$

Lower bound 2: This bound takes advantage of the lexicographical sorting of tasks, which ensures that unassigned tasks at level i address bay l_i or bays above. Consider two QCs k and v where k operates above v. Let c^k and c^v denote the corresponding estimated finishing times with $c^k > c^v$. Since the remaining unassigned tasks belong to bays located above the position of QC k, QC v must remain idle as long as QC k has not finished its service. This means that QC v is blocked for a time period of length $c^k - c^v$. More precisely, the time QC v is blocked by QC k is given by $b_v^k = \max\{w^k + d^k - c^v, 0\}$, which allows for the fact that late ready times do not effect blocking. For a node at level i of the search tree, the expected total occupation time of the cranes including time periods in which they are blocked by QC k is

$$o^k = \sum_{v=1}^{q} c^v + \sum_{v=1}^{k-1} b_v^k + \sum_{j=i+1}^{n} p_j. \tag{7.32}$$

It is calculated as the sum of the QC finishing times at the current state of task assignment plus the total blocking time and the remaining workload of tasks unassigned so far. The shortest possible makespan results if the total occupation time is uniformly distributed to the cranes. This leads to the second lower bound

$$LB_2 = \max_{k \in Q}\left\{\frac{o^k}{q}\right\}. \tag{7.33}$$

Lower bound 3: To obtain a third lower bound, a partial schedule for the given partial task-to-QC assignment is computed and its makespan is determined as described in the subsequent Sects. 7.2.3 and 7.2.4. This makespan serves as LB_3. Obviously, LB_3 dominates LB_1 but the incurred computational cost is comparably high. Therefore, LB_3 is computed only if LB_1 and LB_2 did not effect a bounding. If a partial schedule is completed by the last task n and $LB_3 < UB$ holds, the schedule represents a new best solution. In this case UB is updated by LB_3.

Branching: To continue the partial task-to-QC assignment, nodes passing the bounding criteria are branched by adding successor nodes to the search tree at level $i+1$. The successor nodes represent possible assignments $m_{i+1} = k$ of task

$i+1$ to one of the QCs $k \in Q$. They must meet two branching criteria, which limit the tree search to the inspection of promising task-to-QC assignments for which a unidirectional schedule can be created.

Branching criterion 1: Unidirectional schedules show no change in the direction of QC movement after the initial crane repositionings. The first branching criterion prohibits task-to-QC assignments that lead to unavoidable changes in the direction of movement. Such a change is inevitable for precedence constrained tasks of the same bay if the QC of the succeeding task operates above the QC of the preceding task. Formally, for precedence constrained task pairs $(j, i+1) \in \Phi$ the following condition must hold

$$m_j \geq m_{i+1}. \tag{7.34}$$

The existence of precedence constraints leads to a strong reduction of branches in the tree search. Note that task $i+1$ is assigned to QC 1 if task j is assigned to QC 1. Task $i+1$ is assigned either to QC 1 or to QC 2 if task j is assigned to QC 2. Generally, task $i+1$ is assigned to one of the QCs $1, 2, \ldots, m_j$. For this reason, at tree level i, at most m_j nodes are created for further inspection.

Branching criterion 2: With the only exception of the initial repositioning, QCs are not allowed to move downward in a unidirectional schedule. However, repositioning QC v downward does not make sense if another QC w can reach the bay position of the considered task $i+1$ earlier, because w will stay idle while v processes task $i+1$. Therefore, if adjacent QCs v and $w = v-1$ are not involved in the partial task-to-QC assignment, a successor node $m_{i+1} = v$ is only added to the search tree, if the following condition holds:

$$r^v + t^v_{0,i+1} \leq r^w + t^w_{0,i+1}, \tag{7.35}$$

As an example consider the instance in Fig. 7.4 but assume that task 1 is assigned to the upper QC ($v = 2$). This forces the lower QC ($w = 1$) to move downward and stay idle. Since there are no ready times given for both cranes and QC v has a longer travel time to reach bay 1, Condition (7.35) does not hold, which prevents a further investigation of this partial task-to-QC assignment.

7.2.3 Sequencing of Tasks

After the tasks have been assigned to QCs, a task sequence is determined for each crane. Although a feasible schedule can be derived from any combination of QC task sequences, only one certain combination of sequences leads to a unidirectional schedule. Due to the lexicographical sorting, the right sequences are already determined by the order in which the tasks have been treated in the assignment process. In other words, the first task assigned to a QC is the first in its sequence, the next assigned task is the second in its sequence, and so on. Each change in a

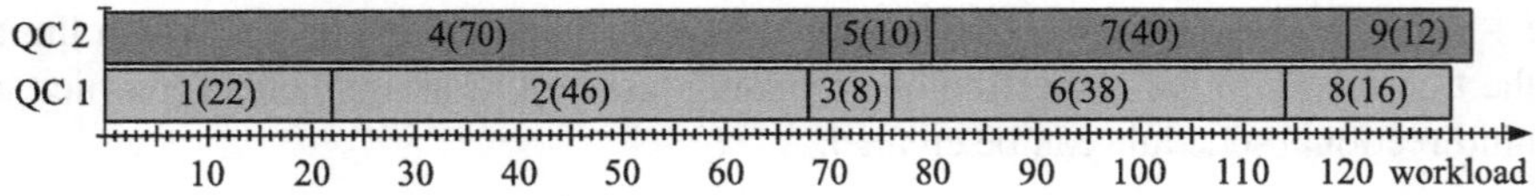

Fig. 7.8 Task sequences for the task-to-QC assignment of Fig. 7.7

sequence means that a unidirectional schedule can no longer be built. Precedence constrained tasks can obviously not be changed in sequence. Changing the sequence of unconstrained tasks belonging to different bays inevitably requires a change in the moving direction of the crane. Consequently the sequencing of tasks follows the lexicographical sorting which entails no computational effort for the UDS heuristic.

Example 7.2: Sequencing of tasks

From the task-to-QC assignment shown in Fig. 7.7 one obtains task sequences 1-2-3-6-8 and 4-5-7-9 for QCs 1 and 2, respectively. The sequences and the resulting distribution of the workload among the two cranes are shown in Fig. 7.8.

7.2.4 Scheduling of Tasks

Generally, one can build different schedules from a set of QC task sequences depending on the priority given to tasks which are forbidden to be processed simultaneously. Therefore, a schedule generation scheme is applied to assign priorities to tasks. In order to generate non-delay crane schedules, Kim and Park (2004) apply the list scheduling scheme. In a non-delay schedule, cranes do not remain idle while they could start processing a task. Since the set of non-delay schedules does not necessarily contain the optimal solutions, the approach conducts a heuristic reduction of the search space. Lim et al. (2007) generate unidirectional schedules by scheduling tasks in the order of increasing bay position, i.e., priority is given to the task with lower bay position. This schedule generation scheme can, however, not be applied to QCSP formulations that respect a safety margin between QCs. A safety margin can necessitate to give priority to a task with a higher bay position in order to generate an optimal unidirectional schedule (see tasks 3 and 4 in the schedule shown in Fig. 7.4 on page 94).

A schedule generation scheme capable of capturing the optimal QCSP schedule is based on the disjunctive graph model, which is well known in the field of machine scheduling. In the approach of Sammarra et al. (2007) the sequencing and the scheduling of QC tasks are commonly based on a problem representation using disjunctive graphs.

In the study at hand the disjunctive graph model is applied to build a unidirectional schedule for every task-to-QC assignment generated in the tree search. In face of a large number of possible task-to-QC assignments, the model turns out to be very valuable because it reveals an efficient way for incorporating crane interference issues into the schedule generation scheme.

In the disjunctive graph model, all tasks $i \in \overline{\Omega}$ are represented as nodes. The task sequence of crane $k \in Q$ is represented by a set of conjunctive arcs A_k which defines a path from node 0 to node T. The precedence constrained task pairs of set Φ are represented by a further set of conjunctive arcs A_Φ. The set of all conjunctive arcs is defined as $A = \bigcup_{k \in Q} A_k \cup A_\Phi$. The further task pairs that are forbidden to be processed simultaneously are represented by pairs of disjunctive arcs. A pair of disjunctive arcs between tasks expresses the two possible orders of processing them. The set of disjunctive arcs in the graph is denoted as D. It contains arcs for pairs of non-simultaneous tasks defined in Ψ. Additionally, D contains arcs for task pairs which are not precedence constrained but can cause crane interference under the given task-to-QC assignment. This latter task set is defined as $\Psi_\Theta = \{(i,j) \in \Omega^2 \backslash \Phi \mid (i,j,m_i,m_j) \in \Theta\}$. Now, the set of all disjunctive arcs is $D = \{(i,j) \in \Omega^2 \mid (i,j) \in \Psi \cup \Psi_\Theta \vee (j,i) \in \Psi \cup \Psi_\Theta\}$. Note that if tasks i and j cannot be processed simultaneously, both arcs, (i,j) and (j,i), must enter D.

Weights are defined for conjunctive and disjunctive arcs in different ways. Weights for the conjunctive arcs $(i,j) \in \bigcup_{k \in Q} A_k$ are given by

$$w_{ij} = \begin{cases} r^{m_j} + t_{0j}^{m_j}, & \text{if } i = 0 \text{ and } j \in \Omega, \\ p_i, & \text{if } i \in \Omega \text{ and } j = T, \\ p_i + t_{ij}, & \text{otherwise.} \end{cases} \tag{7.36}$$

These weights assess a processing time of a task or a ready time of a crane plus the travel time needed by a QC to move to the bay position of the next task. Weights for the conjunctive arcs $(i,j) \in A_\Phi \backslash \bigcup_{k \in Q} A_k$ which belong to precedence constraints not contained in the task sequences and weights for the disjunctive arcs $(i,j) \in D$ are defined slightly differently by

$$w_{ij} = p_i + \Delta_{ij}^{m_i m_j}. \tag{7.37}$$

Next to a task processing time these weights also reflect the required temporal distance for a safe crane movement, before the next task can be started. Summarizing, the disjunctive graph, which is obtained from a task-to-QC assignment, is denoted as $G = (\overline{\Omega}, A, D, W)$, where $W = [w_{ij}]$ represents the arc weights.

From the disjunctive graph G of a scheduling problem one can obtain a feasible schedule by selecting one arc of each pair of disjunctive arcs (and dropping the other) such that the resulting graph G' becomes acyclic. The unidirectional schedule is derived from G by always selecting those arcs from the pairs of disjunctive arcs that are directed from nodes of the upper QC toward nodes of the lower QC. This means that whenever two tasks cannot be processed simultaneously, the schedule generation scheme gives *priority to the upper QC*. Consequently, a cycle in G' can be effected only by arcs from A_Φ. However, due to the first branching criterion (7.34), arcs corresponding to precedence constraints are strictly downward oriented too and, therefore, G' cannot become cyclic. The makespan of the resulting unidirectional schedule is computed as the length of the longest path in G'.

Example 7.3: Scheduling of QC tasks (continued Example 7.2)

The schedule generation is illustrated for the task-to-QC assignment shown in Fig. 7.7 and the corresponding task sequences shown in Fig. 7.8. Using the assignment and the task sequences, the node and arc sets shown in Table 7.3 are generated as described above. Here, A_1 and A_2 are the arc sets representing the task sequences of QCs 1 and 2, respectively. A_Φ represents the precedence constraint between tasks 1 and 2 and between tasks 5 and 6. The disjunctive arcs in D are used for handling crane interference.

The disjunctive graph G is shown in Fig. 7.9. The arc weights are calculated using (7.36) and (7.37). The composition of weights is shown in detail for arcs $(0,4),(4,5),\ (7,8)$, and $(9,T)$ in the figure.

Giving priority to the upper QC whenever two tasks cannot be processed simultaneously means selecting the downward oriented arc from each of the five disjunctive arc pairs of graph G. Figure 7.10 shows the directed graph G' that is obtained from G. The corresponding longest path $(0,4,5,7,8,T)$ is of length 145. This value measures the makespan of the schedule shown in Fig. 7.4. While a two-crane problem has been considered for illustrating the proposed heuristic, the procedure can be applied without restrictions to generate unidirectional schedules for larger problems.

Table 7.3 Objects for the construction of the disjunctive graph

$$
\begin{aligned}
\overline{\Omega} &= \{0,1,2,3,4,5,6,7,8,9,T\}\\
A_1 &= \{(0,1),(1,2),(2,3),(3,6),(6,8),(8,T)\}\\
A_2 &= \{(0,4),(4,5),(5,7),(7,9),(9,T)\}\\
A_\Phi &= \{(1,2),(5,6)\}\\
D &= \{(3,4),(4,3),(4,6),(6,4),(4,8),(8,4),(5,8),(8,5),(7,8),(8,7)\}
\end{aligned}
$$

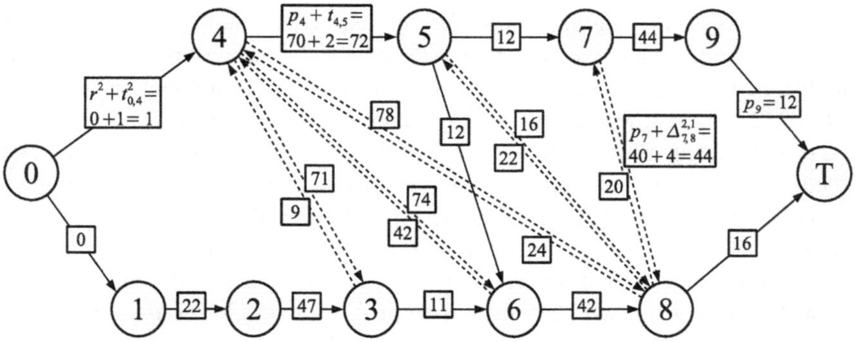

Fig. 7.9 Graph G obtained for the task-to-QC assignment of Fig. 7.7

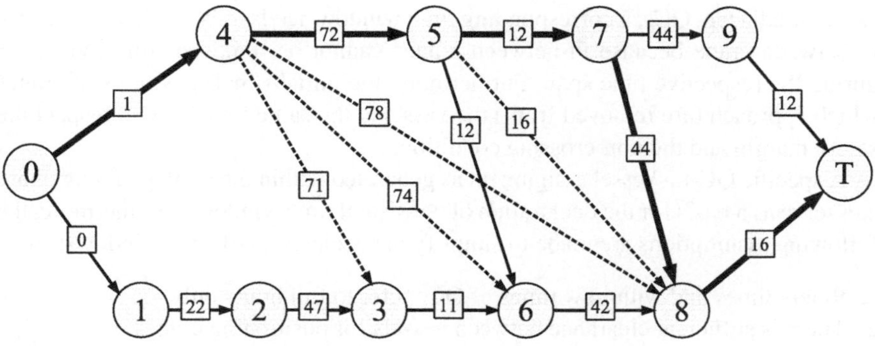

Fig. 7.10 Graph G' corresponding to Fig. 7.9

7.3 The QCSP with Time Windows

The computational study of the BACAP has revealed that the quality of berth plans is improved if variable-in-time QC-to-Vessel assignments are considered. While the above QCSP formulation can handle non-zero ready times of cranes, it cannot be applied if QCs are (temporarily) removed from a vessel during the service. Hence, the incorporation of *time windows for the cranes* becomes necessary. A time window defines a time span at which a crane is available at the vessel. The resulting problem is called the *Quay Crane Scheduling Problem with Time Windows* (QCSPTW). To access the QCSPTW a consistent declaration of time windows for cranes is discussed, followed by a mathematical formulation of the problem. Afterwards, an adaptation of the UDS heuristic is presented to solve the QCSPTW.

7.3.1 Declaration of Time Windows for Cranes

Since cranes can be temporarily removed from a vessel during the service, a crane can possess multiple time windows. The following notation is used to state time windows for a crane $k \in Q$. Let $TW_k = \{1,\ldots,\tau_k\}$ denote the set of time windows of k where τ_k is the number of time windows.

Each time window $u \in TW_k$ is defined by the quadruple $(r^{ku}, d^{ku}, l_0^{ku}, l_T^{ku})$ with:

- The ready time r^{ku}, i.e., the begin of the time window
- The withdraw time d^{ku}, i.e., the end of the time window
- The initial crane position l_0^{ku} at the begin of the time window
- The final crane position l_T^{ku} at the end of the time window

Without loss of generality it is assumed that the time windows of a crane are indexed according to increasing ready times, i.e., $r^{k,1} < r^{k,2} < \cdots < r^{k,\tau_k}$.

In order to obtain feasible QCSPTW solutions, time windows for cranes must be consistently declared. For example, if overlapping time windows are declared for two non-adjacent QCs, a corresponding time window needs to be declared for every in-between crane because in-between cranes cannot be removed from the vessel during the respective time span. Furthermore, the initial (final) positions of cranes which approach (are removed from) the vessel at the same time have to respect the safety margin and the non-crossing condition.

A specific QC-to-Vessel assignment as generated within the berth planning phase can serve as a basis for the declaration of consistent time windows. Furthermore, the following assumptions are made to simplify the declaration of time windows:

1. Ready times and withdraw times of QCs refer to full hours only.
2. There is sufficient clearance between vessels for positioning cranes.
3. The time needed to move a QC from one vessel to another vessel is neglected.
4. QCs which serve the considered vessel in its last handling hour (according to the given QC-to-Vessel assignment) stay until the service is completed.

The first assumption is justified because the BACAP assigns cranes to vessels on an hourly basis. The second assumption ensures that approaching cranes and removing cranes can be positioned outside of the vessel area without getting into conflict with cranes serving other vessels. The third assumption has already been stated for the BACAP. This simplification allows to focus on the considered vessel without taking into account the origin of an approaching QC or the destination of a removed QC. The fourth assumption ensures that there is always a feasible solution for a QCSPTW instance existing under any QC-to-Vessel assignment. For this purpose the latest-starting time window τ_k of those cranes $k \in Q$ which are assigned to the vessel in its last handling hour receives an infinite withdraw time. Such time windows are called *open-ended* time windows. An open-ended time window enables a QC to remain at the vessel and complete handling operations. It is mathematically described by $d^{k,\tau_k} = M$.

Following these assumptions *ready times* and *withdraw times* of cranes can be derived straightforward from a specific QC-to-Vessel assignment of a BACAP solution. Merely a time transformation is required. The berthing time of the considered vessel in the BACAP is transformed into time 0 in the QCSPTW while the time unit is changed from hours to minutes. This transformation is illustrated in Fig. 7.11 where a single vessel berths at time $s_1 = 3$ and departs at time $e_1 = 9$. Consider QC 1, which approaches the vessel 1 h after the berthing time. The crane is removed from the vessel 2 h later. Hence, the corresponding time window $u = 1$ of QC $k = 1$ in the QCSPTW shows a crane ready time $r^{1,1} = 60$ and a crane withdraw time $d^{1,1} = 180$.

The *initial crane positions* of time windows are determined differently depending on the crane ready time. Cranes which are assigned to the vessel at its berthing time are lined up alongside the vessel with inter-crane clearance as required from the safety margin. Cranes which approach the vessel at a later point in time are initially positioned outside of the vessel area. For a vessel with b bays, these outside positions can be addressed by virtual bays $0, -1, \ldots$ and $b+1, b+2, \ldots$, respectively. Assume that a QC v approaches a vessel while the service is running. Let w_{lo} and w_{up} denote

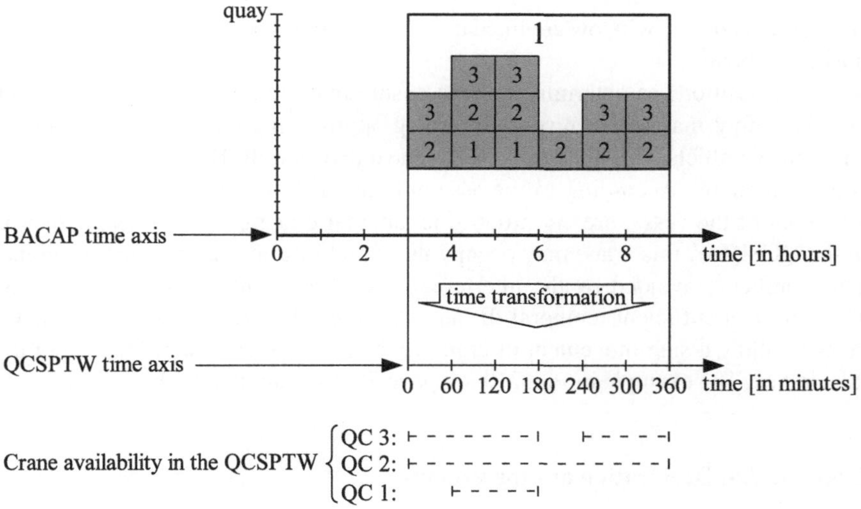

Fig. 7.11 Time transformation between BACAP and QCSPTW

the downmost crane and the upmost crane among the already assigned QCs. The initial position of crane v at the begin of time window $u \in TW_v$ is given by

$$l_0^{v,u} = \begin{cases} 1 - \delta_{v,w_{\text{lo}}}, & \text{if } v < w_{\text{lo}}, \\ b + \delta_{v,w_{\text{up}}}, & \text{if } v > w_{\text{up}}. \end{cases} \tag{7.38}$$

Here $\delta_{v,w_{\text{lo}}}$ ($\delta_{v,w_{\text{up}}}$) denotes the smallest allowed difference between the bay positions of QCs v and w_{lo} (w_{up}) as calculated by (7.7) on page 89. Eq. (7.38) positions a crane at the nearest feasible position outside of the vessel area. The first case of (7.38) applies if v is a crane positioned below the already assigned cranes at the quay. It can be seen that v is initially positioned below bay 1. For example, if the safety margin is set to $\delta = 1$ bay, and v is the crane adjacent to w_{lo} ($v = w_{\text{lo}} - 1$), crane v is initially positioned at bay $l_0^{v,u} = 1 - \delta_{v,w_{\text{lo}}} = 1 - 2 = -1$. If the next crane $v' = w_{\text{lo}} - 2$ approaches the vessel at the same time, the initial position of v' is set to bay $l_0^{v',u'} = 1 - \delta_{v',w_{\text{lo}}} = 1 - 4 = -3$, and so on. The second case of (7.38) applies if v is a crane positioned above the already assigned cranes at the quay. In this case the initial position is set to a bay position above bay b.

For calculating *final crane positions* of time windows, (7.38) is modified by replacing $l_0^{v,u}$ by $l_T^{v,u}$. Furthermore, v refers to the QC that is removed from the vessel, and w_{lo} (w_{up}) refers to the downmost (upmost) QC among the cranes that remain at the vessel. From these modifications (7.38) calculates the final position of a removed crane at the end of its time window. Note that the equation cannot be applied to open-ended time windows because w_{lo} and w_{up} are not defined at the end of a vessel's service. Since the final crane positions of open-ended time windows have no impact on the makespan of a schedule, they are of no particular interest within the QCSPTW. For reasons of completeness the final positions of cranes with an open-ended time window are lined up alongside the vessel as described for initial positions above.

Crane positions as determined above ensure that the non-crossing requirement and the safety margin are respected among approaching and removing cranes at any time. Conflicts with operating QCs are avoided as well. However, one potential conflict remains. According to the assumptions made for the QCSP, a crane can move out of the vessel area in order to let another crane process a task. However, in the QCSPTW, this crane may occupy the initial position of approaching cranes. This conflict is avoided by the interference constraints introduced for the QCSP. They insert a sufficient temporal distance between the processing of consecutive tasks to allow a safe movement of cranes, which is also sufficient for temporarily idle cranes. The example in Fig. 7.2b on page 88 gives an idea of this effect.

Example 7.4: Declaration of time windows

The declaration of time windows for cranes is demonstrated using the QC-to-Vessel assignment shown in Fig. 7.11. The vessel is assumed to have $b = 10$ bays and the safety margin is set to $\delta = 1$ bay. The following time windows are derived from the depicted QC-to-Vessel assignment:

QC 1 is assigned to the vessel from the beginning of the second service period to the end of the third service period. According to the time transformation, the corresponding time window in the QCSPTW shows a ready time of $r^{1,1} = 60$ and a crane withdraw time of $d^{1,1} = 180$. The initial position of the crane at the begin of the time window is calculated using (7.38) as $l_0^{1,1} = 1 - \delta_{1,2} = -1$, i.e., the crane is placed outside of the vessel area. The calculation takes into account that QC 2 is already assigned to the vessel at the ready time. Also the final position is set to $l_T^{1,1} = 1 - \delta_{1,2} = -1$ because QC 2 remains at the vessel while QC 1 is removed. Hence, one time window $(r^{1,1}, d^{1,1}, l_0^{1,1}, l_T^{1,1}) = (60, 180, -1, -1)$ is declared for QC 1.

QC 2 is assigned to the vessel throughout its entire handling time. A time window $(r^{2,1}, d^{2,1}, l_0^{2,1}, l_T^{2,1}) = (0, M, 1, 1)$ is declared for a consistent treatment. Note that this time window is open-ended as stated above by the fourth assumption. The initial position of the crane is taken from the lining up of those cranes that are assigned to the vessel at its berthing time. The final position is taken from lining up all cranes with an open-ended time window alongside the vessel.

The assignment of QC 3 to the vessel requires two time windows. The first time window is defined by $(r^{3,1}, d^{3,1}, l_0^{3,1}, l_T^{3,1}) = (0, 180, 3, 12)$. Here, the initial position follows from lining next to QC 2 with respect to the safety margin. The final position is calculated by $b + \delta_{3,2} = 12$ which places the QC outside of the vessel area and, thereby, respects that QC 2 remains at the vessel. The second time window of QC 3 is $(r^{3,2}, d^{3,2}, l_0^{3,2}, l_T^{3,2}) = (240, M, 12, 3)$.

7.3.2 Optimization Model

A mathematical formulation of the QCSPTW is derived by extending the QCSP formulation given in Sect. 7.1. The formulation uses the additional terms $t_{0i}^{ku} = \hat{t}|l_0^{ku} - l_i|$ and $t_{iT}^{ku} = \hat{t}|l_T^{ku} - l_i|$. The former denotes the travel time of QC k between the initial crane position of time window $u \in TW_k$ and the bay position of task $i \in \Omega$. The latter denotes the travel time between the position of task i and the final crane position at the end of the time window u. Moreover, binary decision variables y_i^{ku} are introduced, set to 1 if and only if task i is processed by QC k in its time window u.

The QCSPTW is formulated as follows:

$$\text{minimize } c_T \tag{7.39}$$

subject to

$$\sum_{u \in TW_k} y_i^{ku} = \sum_{j \in \Omega^0} x_{ji}^k \qquad \forall i \in \Omega, \forall k \in Q, \tag{7.40}$$

$$c_i - p_i \geq y_i^{ku}(r^{ku} + t_{0i}^{ku}) \qquad \forall i \in \Omega, \forall k \in Q, \forall u \in TW_k, \tag{7.41}$$

$$c_i \leq M(1 - y_i^{ku}) + d^{ku} - t_{iT}^{ku} \qquad \forall i \in \Omega, \forall k \in Q, \forall u \in TW_k, \tag{7.42}$$

$$y_i^{ku} \in \{0,1\} \qquad \forall i \in \Omega, \forall k \in Q, \forall u \in TW_k, \tag{7.43}$$

and (7.14)–(7.25), (7.27)–(7.29).

As in the QCSP, the objective pursues the minimization of the makespan of the schedule. Constraints (7.40) ensure that every task is processed within one time window of the assigned crane. Constraints (7.41) ensure that a task is not started earlier than the crane ready time of the addressed time window plus the time needed by the crane to move from its initial position to the task. Constraints (7.42) ensure that each crane is able to reach its final position at the withdraw time of a time window. The model is completed by the Constraints (7.14)–(7.25) and (7.27)–(7.29). The QCSPTW is classified by $Group, prec \mid TW, pos, move \mid cross, save \mid max(compl)$.

7.3.3 Adaptation of the UDS Heuristic

The UDS heuristic is adapted for the QCSPTW, referred to as the *UDSTW heuristic*. The following adaptations are necessary:

1. The task-to-QC assignments generated by the *S-TASKS* and the *S-LOAD* rule may not allow for the generation of feasible schedules in the presence of time windows. In this case both rules are repeated, but this time only cranes with an open-ended time window are considered. The derived task-to-QC assignments lead to feasible schedules. The obtained schedules yield the initial upper bound for the UDSTW heuristic.

2. The lower bounds that have been introduced for the UDS heuristic already respect ready times r^k and initial positions l_0^k of QCs. The bounds are applicable in the UDSTW heuristic by replacing these values with the corresponding values of the earliest time windows of the cranes, namely $r^{k,1}$ and $l_0^{k,1}$.
3. The second branching criterion enforces that a task is assigned to the QC that can reach it in the fastest possible way. However, if the chosen crane shows no time window sufficiently large to process the task, the QCSPTW cannot be solved feasible. For this reason, the second branching criterion is disabled in the UDSTW heuristic.
4. The scheduling rule derived from the disjunctive graph representation of the QCSP is applicable within the UDSTW heuristic. However, it may happen that (i) the derived earliest starting time of a task does not fall within a time window of the assigned QC, or (ii) the task cannot be completed before the addressed time window ends. In both cases the task is postponed by shifting it to the next time window of the assigned crane that is sufficiently large to allow processing the task and the required crane movement. Successor tasks and tasks assigned to QCs with lower priority are postponed accordingly. If no appropriate time window can be found for a task, the task-to-QC assignment does not lead to a feasible unidirectional schedule. The assignment is not further investigated by the UDSTW heuristic.

The UDSTW heuristic searches the space of unidirectional schedules of a QCSPTW instance. However, in the presence of time windows for the cranes, this search space reduction can exclude the optimal schedule even if no precedence relations exists among tasks. Figure 7.12 shows an optimal solution for a QCSPTW instance with two cranes assigned to the vessel. In this example, the first assumption stated in Sect. 7.3.1 is dropped, i.e., ready times and withdraw times of QCs do not refer to full hours in this example. QC 1 is available within two time windows $(r^{1,1}, d^{1,1}, l_0^{1,1}, l_T^{1,1}) = (0, 15, 1, -1)$ and $(r^{1,2}, d^{1,2}, l_0^{1,2}, l_T^{1,2}) = (25, M, -1, 1)$. QC 2 is assigned to the vessel during the entire service interval, which is represented by the single open-ended time window $(r^{2,1}, d^{2,1}, l_0^{2,1}, l_T^{2,1}) = (0, M, 3, 3)$. In order to capture the optimal solution with a makespan of 69 time units, QC 1 needs to process task 2 within its first time window and tasks 1 and 3 within its second time

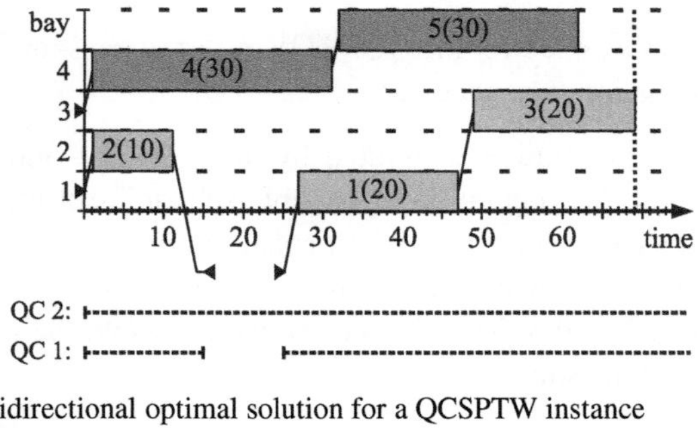

Fig. 7.12 A non-unidirectional optimal solution for a QCSPTW instance

window. However, this task sequence contradicts the lexicographical task indexing, i.e., the resulting schedule is not a unidirectional one. It can therefore not be generated by the UDSTW heuristic. The example illustrates that the reduction of the solution space to unidirectional schedules might have a stronger impact on the quality of solutions for the QCSPTW than for the QCSP. A quantitative investigation on the performance of the heuristic is left to the computational study in the subsequent section.

7.4 Computational Study

The following tests assess the competitiveness of the UDS heuristic against methods published in the literature, the sensitivity of the solutions on the problem parameter settings, and the performance of the UDSTW in solving the problem variant with time windows for the cranes:

- Performance of the UDS heuristic:

  Test 7.1: Performance on small QCSP instances
  Test 7.2: Performance on large QCSP instances
  Test 7.3: Solution quality and runtime demand

- Sensitivity on problem parameter settings:

  Test 7.4: Sensitivity on the task definition
  Test 7.5: Sensitivity on the safety margin

- Performance of the UDSTW heuristic:

  Test 7.6: Solution quality and runtime demand of the UDSTW heuristic

The heuristics have been implemented in JAVA. A PC P4 2.8 GHz is used for the tests.

Benchmark Instances

A suite of benchmark problems is used that has been introduced by Kim and Park (2004). It contains nine instance sets of different problem size with ten instances each, see Table 7.4. For each instance the order of processing the tasks of a bay is completely determined by precedence constraints, i.e., a unique lexicographical

Table 7.4 QCSP benchmarks of Kim and Park (2004)

	Set								
	A	B	C	D	E	F	G	H	I
# Tasks n	10	15	20	25	30	35	40	45	50
# QCs q	2	2	3	3	4	4	5	5	6

sorting of tasks is always possible. The QC ready times r^k are zero in all instances. The safety margin δ is set to one bay. The travel time $\hat{t}$ of QCs is set to one time unit per bay. Variations in these settings are stated in the respective tests.

Test 7.1: Performance on small QCSP instances

This test assesses the performance of the UDS heuristic and compares it with the QCSP solution methods published in the literature, see Table 7.5.

Since the studies of Kim and Park (2004), Moccia et al. (2006), and Sammarra et al. (2007) tackled only the first 37 instances, named $k13$ to $k49$, from the benchmark suite, the comparison is carried out on this subset of benchmarks. Moccia et al. (2006) report optimal solutions for 28 of these instances and tight lower bounds for the remaining nine instances. While it is unknown whether the solutions are feasible with respect to the corrected interference constraints, they can still serve as lower bounds. This holds because optimal solutions generated for a model with incomplete interference constraints will show a makespan that is not larger than the minimum makespan of the corrected model's solutions.

Table 7.6 shows the computational results of this first experiment. The optimal makespan or best known lower bound to each of the instances appears in column f_{opt}. Note that previous studies weighted the obtained makespan of a schedule by a fixed factor of 3, as is also done here to ease comparison. The performance of the methods is reported on the basis of the relative error RE in percent of the best found solution f_{best} against f_{opt}, i.e., $RE = (f_{\text{best}} - f_{\text{opt}})/f_{\text{opt}} \times 100$. For the UDS heuristic the objective function value of the best found solution is reported in cases where it fails to reach f_{opt}.

With the exception of the algorithms of Kim and Park (2004) all compared methods solve instance set A to optimality. The observed deviation between f_{best} and f_{opt} for the UDS solution of $k22$ results from the interference constraints (7.10)–(7.12) in the corrected model. Replacing these constraints by (7.2), (7.4), and (7.6), which are used by Sammarra et al. (2007), the UDS heuristic always reaches f_{opt}. For this reason it is assumed that the corrected solution for $k22$ is optimal with respect to the revised QCSP model. For instance sets B and C, the relative error of the Tabu Search clearly increases against the Branch-and-Cut algorithm and the UDS heuristic. The Branch-and-Cut algorithm and UDS generate schedules of identical quality. The relative error of 2.26% for the UDS solution of $k42$ stems again from the strict

Table 7.5 QCSP solution methods published in the literature

Abbr.	Method	Reference
B&B	Branch-and-bound	Kim and Park (2004)
GRASP	Greedy randomized adaptive search procedure	Kim and Park (2004)
B&C	Branch-and-cut	Moccia et al. (2006)
TS	Tabu search	Sammarra et al. (2007)

Table 7.6 Performance comparison of QCSP solution methods (*RE* in percent)

No.	Set	f_{opt}	B&B	GRASP	B&C	TS	UDS	f_{best}
*k*13	A	453	0.00	0.00	0.00	0.00	0.00	
*k*14	A	546	0.00	0.00	0.00	0.00	0.00	
*k*15	A	513	0.00	0.58	0.00	0.00	0.00	
*k*16	A	312	2.88	2.88	0.00	0.00	0.00	
*k*17	A	453	0.66	0.66	0.00	0.00	0.00	
*k*18	A	375	0.00	0.00	0.00	0.00	0.00	
*k*19	A	543	1.66	1.66	0.00	0.00	0.00	
*k*20	A	399	20.30	20.30	0.00	0.00	0.00	
*k*21	A	465	0.00	0.00	0.00	0.00	0.00	
*k*22	A	537	34.08	34.08	0.00	0.00	0.56	540[b]
*k*23	B	576	0.00	2.60	0.00	1.04	0.00	
*k*24	B	666	0.45	1.35	0.00	0.45	0.00	
*k*25	B	738	0.00	0.41	0.00	0.41	0.00	
*k*26	B	639	0.00	1.88	0.00	0.00	0.00	
*k*27	B	657	0.00	4.57	0.00	0.46	0.00	
*k*28	B	531	1.13	3.39	0.00	0.00	0.00	
*k*29	B	807	0.00	1.49	0.00	0.37	0.00	
*k*30	B	891	0.00	1.68	0.00	0.00	0.00	
*k*31	B	570	0.00	0.00	0.00	0.00	0.00	
*k*32	B	591	0.00	1.02	0.00	0.00	0.00	
*k*33	C	603	0.00	10.45	0.00	0.00	0.00	
*k*34	C	717	0.00	6.28	0.00	2.51	0.00	
*k*35	C	684	0.88	2.19	0.00	0.88	0.00	
*k*36	C	678	6.19	4.42	0.00	0.44	0.00	
*k*37	C	510	1.18	5.88	0.00	1.76	0.00	
*k*38	C	613.67[a]	3.15	7.55	0.71	0.71	0.71	618
*k*39	C	508.38[a]	8.58	13.89	0.91	2.09	0.91	513
*k*40	C	564	2.13	5.85	0.00	0.53	0.00	
*k*41	C	585.06[a]	11.78	9.73	0.50	1.53	0.50	588
*k*42	C	560.31[a]	4.94	18.86	1.73	2.80	2.26	573[b]
*k*43	D	859.32[a]	10.67	9.62	4.38	2.29	1.94	876[c]
*k*44	D	820.35[a]	7.15	4.59	0.20	1.66	0.20	822
*k*45	D	824.88[a]	4.38	5.83	1.83	3.29	1.11	834[c]
*k*46	D	690	2.61	6.52	0.00	0.00	0.00	
*k*47	D	792	15.15	1.89	0.00	0.00	0.00	
*k*48	D	628.87[a]	6.38	6.38	2.56	5.43	1.61	639[c]
*k*49	D	879.22[a]	4.07	10.55	5.43	3.73	1.68	894[c]
ARE (%)			4.06	5.65	0.49	0.87	0.31	

[a]Lower bound; [b]Corrected solution; [c]New best solution

interference handling. The UDS heuristic returns the Branch-and-Cut solution if the new interference constraints are not applied. For the larger instances of set D, the Branch-and-Cut algorithm often fails to reach the optimum within the allowed runtime of 2 h. Here, the UDS heuristic is clearly superior to all other methods. For instances *k*43, *k*45, *k*48, and *k*49 it delivers new best solutions. This is also reflected by a comparison of the average relative error (*ARE*) observed for the heuristics. The

Table 7.7 Runtime comparison of QCSP solution methods (average-in-set in minutes)

Set	B&B	GRASP	B&C	TS	UDS
A	0.44	0.35	1.01	1.52	1.12×10^{-5}
B	17.53	1.46	8.91	5.86	3.68×10^{-5}
C	564.47	3.16	72.19	21.75	6.26×10^{-4}
D	809.73	7.56	102.49	48.68	3.43×10^{-3}

achieved excellent solution quality implies that at least the smaller instances of Kim and Park (2004) have optimal solutions which are unidirectional, too. In total the UDS heuristic is capable of solving all instances to optimality or to the best solution quality known so far.

The average computation time demand of the various algorithms (as reported in the literature) is presented for each of the four instance sets in Table 7.7. The machines used were a PC P2 466 MHz for the Branch-and-Bound method and the GRASP heuristic of Kim and Park, a PC P4 2.5 GHz for the Branch-and-Cut method of Moccia et al., a PC P4 2.66 GHz for the Tabu Search of Sammarra et al., and a PC P4 2.8 GHz for the UDS heuristic. Although within milliseconds, the computation times of the UDS procedure are specified in minutes for the purpose of comparability. Despite the fact that it has been tested on the fastest machine, it can be seen that the UDS heuristic tremendously cuts down the computation times.

Test 7.2: Performance on large QCSP instances

The performance of the UDS heuristic on the instance sets A to D encourages one to tackle the larger instances as provided in sets E to I of the benchmark suite, see Table 7.4. They contain problems with up to six QCs and 50 tasks as observed for large container vessels. It is supposed that previous studies had not tackled these problems because the proposed methods ran into their boundaries. Computational results obtained from the UDS heuristic for instances $k50$ to $k102$ are shown in Table 7.8. The UDS heuristic is given a runtime limit of 1 h. Recall from Sect. 7.2.1 that the procedure searches consecutively for unidirectional schedules with respect to upward and downward movements of the QCs. To ensure a fair allocation of computation time, both search processes are performed concurrently. In the event that one of the processes terminates within 30 min, the remaining computation time is made available to the other process. The reported runtime is the sum of the runtimes spent on searching in the two directions. If the limit of 60 min is exceeded, no runtime is reported. In these cases value f_{best} does not necessarily represent the optimal unidirectional schedule. To assess the quality of UDS solutions a lower bound is required. The lower bounds presented in Sect. 7.2.2 do not serve this purpose because they evaluate given (partial) task-to-QC assignments. Therefore, a lower bound on the makespan of the instances is calculated by solving a relaxed QCSP model given in Appendix C using CPLEX.

Table 7.8 Results of the UDS heuristic for the remaining instances in sets D to I

No.	*LB*	f_{best}	*RE*	*Time*	No.	*LB*	f_{best}	*RE*	*Time*
		D					E		
(see Table 7.6 for *k*43–*k*49)					*k*53	657	717	9.13	–
					*k*54	753	774	2.79	0.02
					*k*55	663	684	3.17	0.01
					*k*56	666	690	3.60	0.22
					*k*57	681	705	3.52	0.24
					*k*58	765	786	2.75	0.17
					*k*59	666	687	3.15	0.01
*k*50	723	741	2.49	<0.01	*k*60	765	783	2.35	0.19
*k*51	777	798	2.70	<0.01	*k*61	618	639	3.40	0.04
*k*52	939	960	2.24	<0.01	*k*62	828	837	1.09	0.01
ARE (%)			2.48		*ARE* (%)			3.50	
		F					G		
*k*63	927	948	2.27	1.51	*k*73	837	870	3.94	31.71
*k*64	714	741	3.78	1.06	*k*74	822	843	2.55	4.71
*k*65	816	837	2.57	1.61	*k*75	657	675	2.74	0.37
*k*66	903	924	2.33	0.63	*k*76	825	852	3.27	0.90
*k*67	858	882	2.80	0.24	*k*77	672	699	4.02	1.27
*k*68	945	963	1.90	0.03	*k*78	621	642	3.38	8.96
*k*69	783	807	3.07	1.40	*k*79	717	744	3.77	1.52
*k*70	936	957	2.24	0.61	*k*80	720	750	4.17	1.28
*k*71	807	834	3.35	3.77	*k*81	705	738	4.68	1.28
*k*72	720	744	3.33	0.35	*k*82	696	717	3.02	1.03
ARE (%)			2.76		*ARE* (%)			3.55	
		H					I		
*k*83	921	948	2.93	6.37	*k*93	786	816	3.82	–
*k*84	876	897	2.40	3.29	*k*94	765	786	2.75	–
*k*85	945	972	2.86	5.82	*k*95	801	834	4.12	–
*k*86	786	816	3.82	–	*k*96	780	819	5.00	–
*k*87	840	867	3.21	–	*k*97	690	720	4.35	–
*k*88	744	768	3.23	43.73	*k*98	711	735	3.38	23.79
*k*89	822	843	2.55	10.96	*k*99	819	852	4.03	–
*k*90	1,023	1,053	2.93	24.95	*k*100	852	900	5.63	–
*k*91	810	837	3.33	10.74	*k*101	765	813	6.27	–
*k*92	873	897	2.75	34.61	*k*102	870	903	3.79	–
ARE (%)			3.00		*ARE* (%)			4.31	

The gained results of this test show that the UDS heuristic is capable of solving the majority of these instances within the runtime limit. Regarding instance sets E to H, merely three instances are not solved. Set I is the only set for which the heuristic fails in solving the majority of the instances. However, the observed deviation between *LB* and f_{best} ranges moderately within a few percent for all considered instances. This indicates that the UDS heuristic delivers schedules of good quality also for large instances.

Table 7.9 Solution quality at selected runtimes (ARE-in-set in percent)

	Set				
Time	E	F	G	H	I
0	16.00	12.73	19.67	14.77	20.34
1	3.50	2.76	3.75	3.35	4.95
10	3.50	2.76	3.55	3.07	4.60
60	3.50	2.76	3.55	3.00	4.31

Test 7.3: Solution quality and runtime demand

As shown by Test 7.2, the UDS heuristic may not terminate within an acceptable runtime if applied to large-sized instances. For this reason, the relation between runtime and solution quality is investigated here in order to determine a reasonable runtime limit for the heuristic. The test considers the large-sized instances of sets E to I. Table 7.9 reports the *ARE* for each set after running the UDS heuristic for 0, 1, 10, and 60 min. The values shown in the row of time 0 are those of the initial solutions.

As can be seen in the table, the solution quality is drastically improved within the first minute of computation for each of the five instance sets. For sets E and F, no further improvement is observed after that time. This can hardly surprise because most of these instances are solved within less than a minute, but it shows that also for instances with longer runtimes no further improvement takes place. For sets G to I, further improvements are observed. The most improvement, i.e., the largest reduction in the *ARE*, is observed for set I. However, even this *ARE* reduction is only 0.64%. It can be concluded that the UDS heuristic converges very quickly for all instance sets. A runtime limit of 1 min per instance is sufficient to ensure that finding solutions of acceptable quality is at a level of high likelihood.

Test 7.4: Sensitivity on the task definition

According to the QCSP classification scheme of Sect. 4.2.1, defining tasks on the basis of container groups is only one possibility. Two alternatives are to define tasks on the basis of complete bays, e.g., Lim et al. (2007), or on the basis of bay areas, e.g., Winter (1999). The advantage of the container group approach is that a more uniform distribution of workload among QCs can be achieved. However, the QCSP becomes more difficult to solve because a larger number of tasks and precedence relations between pairs of tasks come into the play. This test assesses the dependency of solution quality and computational effort on the different task definitions.

For the test, the QCSP instances and the UDS heuristic are slightly modified. First, within each instance, all tasks belonging to the same bay are combined into a single task. Applying the UDS heuristic to such an instance solves a QCSP with tasks defined on the basis of complete bays. Second, to solve the QCSP with tasks

Table 7.10 Results for different tasks definitions

	Container groups								
Set	A	B	C	D	E	F	G	H	I
ARE (%)	0.06	0.00	0.44	1.40	3.50	2.76	3.55	3.00	4.31
Max. RE (%)	0.56	0.00	2.26	2.70	9.13	3.78	4.68	3.82	6.27
Avg. time	<0.01	<0.01	<0.01	<0.01	6.09	1.12	5.30	26.05	56.38
	Complete bays								
Set	A	B	C	D	E	F	G	H	I
ARE (%)	3.00	2.89	1.77	1.88	4.70	3.73	5.25	3.90	4.78
Max. RE (%)	13.41	15.79	7.08	3.32	14.16	7.06	7.95	6.45	10.98
Avg. time	<0.01	<0.01	<0.01	<0.01	0.86	0.06	0.98	0.88	19.51
	Bay areas								
Set	A	B	C	D	E	F	G	H	I
ARE (%)	5.62	7.06	8.11	7.53	9.79	9.60	14.48	10.53	11.67
Max. RE (%)	19.55	36.32	15.07	12.86	15.53	17.10	23.85	17.04	18.70
Avg. time	<0.01	<0.01	<0.01	<0.01	<0.01	<0.01	<0.01	<0.01	<0.01

defined on the basis of bay areas, a branching criterion is added to the UDS heuristic that forbids task-to-QC assignments which lead to overlapping operation areas of QCs at a vessel.

Table 7.10 shows comprehensive results for the benchmark sets if tasks are defined on the basis of container groups (taken from Tests 7.1 and 7.2), on the basis of complete bays, and on the basis of bay areas. It reports the *ARE* per set, the maximum *RE* observed for an instance within a set, and the required average runtime for solving instances of a set. The relative errors are calculated with respect to the optimal solutions and the lower bounds reported by Moccia et al. (2006) for instances up to *k*49 and with respect to the CPLEX lower bounds presented in Table 7.8 for the instances *k*50 to *k*102.

Comparing the results of tasks defined by container groups with tasks defined by complete bays, one can see that the *ARE*s differ by at minimum 0.47% (set I) and at maximum 2.94% (set A). From these findings the solution quality seems to deteriorate only little if tasks are defined on the basis of complete bays. However, the observed maximum *RE*s differ at a much higher rate. For sets A, B, E, and I the maximum *RE* is even above 10% if tasks are defined by complete bays. Defining tasks by container groups is clearly advantageous for the corresponding instances. The required runtimes decrease significantly if tasks are defined by complete bays. However, since the UDS heuristic can be prematurely terminated after 1 min if tasks are defined by container groups (see Test 7.3), the saving of computation time does not increase the attractiveness of a task definition on the basis of complete bays.

If tasks are defined by bay areas, runtimes are negligible even for the largest instances. However, the solution quality deteriorates drastically compared to

Table 7.11 Relative increase in makespan for different safety margins (average-in-set in percent)

δ	Set									Avg.
	A	B	C	D	E	F	G	H	I	
2	5.75	0.16	0.76	0.20	2.27	0.14	2.56	0.06	0.81	1.41
3	15.04	3.02	8.15	2.16	6.23	4.61	9.16	3.49	3.80	6.18
4	26.98	7.72	19.28	10.63	15.97	15.01	18.29	10.39	9.72	14.89

solutions with tasks defined by container groups. *ARE*s are above 10% for the large instances in sets G, H, and I. The maximum *RE* in set B shows that the makespan of an instance may increase by more than one-third compared to the solution with tasks defined by container groups. Since such an increase in the handling time of a vessel is unacceptable from a vessel operator's point of view, defining tasks by bay areas has to be rejected.

Test 7.5: Sensitivity on the safety margin

This test studies the impact of a safety margin on a schedule. Table 7.11 shows the relative change of the makespan observed in the instance sets for different safety margins against $\delta = 1$. Note, that $\delta = 0$ is not investigated because a QC's uprights occupy the bays adjacent to the crane's location and, therefore, positioning QCs at adjacent bays is technically forbidden as a matter of fact. The derived results confirm that the larger the safety margin is, the more the handling times of vessels increase. While the average increase over all sets is only 1.41% for $\delta = 2$, this value rises to 6.18% and even 14.89% for $\delta = 3$ and $\delta = 4$, respectively. Not surprisingly, the small-sized vessels in instance set A suffer most from a large safety margin, but also large-sized instances show a considerable increase in the makespan. The fluctuation which is observed for a certain value of δ stems from the varying number of bays and QCs involved in the different instance sets, see Table 7.4. The test results indicate that incorporating safety margins in the QCSP model is by no means marginal, it is an increasing need, the more safety requirements grow.

Test 7.6: Solution quality and runtime demand of the UDSTW heuristic

A final test assesses the performance of the UDSTW heuristic in solving crane scheduling instances with time windows for the cranes. For the test, the benchmark instances are modified by declaring time windows for the two upmost QCs as follows. The cranes are assigned to a vessel for the first 2 h of service, removed from the vessel for the following 2 h, reassigned for two more hours, and then finally removed. For instances in sets A and B, removing two QCs from a vessel means to interrupt the service process, which, however, can be dealt with by UDSTW. The UDSTW heuristic is given a runtime limit of 1 h per instance. Table 7.12 shows for

Table 7.12 Results of the UDSTW heuristic

Set	A	B	C	D	E	F	G	H	I
ARE (%)	9.10	8.64	7.04	11.75	7.70	9.44	7.74	9.19	10.35
Max. RE (%)	13.29	12.50	9.96	20.25	11.46	12.87	14.20	12.91	19.46
Avg. time	<0.01	<0.01	0.08	0.83	50.41	60.00	60.00	60.00	60.00

Table 7.13 Solution quality for the QCSPTW at selected runtimes (ARE-in-set in percent)

time	set				
	E	F	G	H	I
0	62.25	64.66	52.10	52.90	46.75
1	7.95	9.80	7.89	9.32	10.68
10	7.76	9.47	7.80	9.27	10.41
60	7.70	9.44	7.74	9.19	10.35

every instance set A to I the *ARE* over the contained instances, the maximum *RE* observed for the instances, and the required average runtime. For calculation of the relative errors, a lower bound on the makespan of QCSPTW instances is derived from solving the mathematical model in Appendix D by CPLEX.

As can be seen from this table, the solutions show considerable relative errors. Although the heuristic terminates within the runtime limit for instances in sets A to D, which means that the best unidirectional schedule has been found, *ARE*s of about 10% and a maximum *RE* of more than 20% (set D) are observed. However, also for the larger instances, where UDSTW systematically fails to terminate within the runtime limit, similar *ARE*s and maximum *RE*s are observed. This means that the solution quality does hardly deteriorate in the problem size.

The fact that the UDSTW heuristic does not terminate within the runtime limit does not necessarily mean that it is unable to find good solutions quickly. For this reason, similar to Test 7.3, the *ARE*s observed for instance sets E to I after running the heuristic for 0, 1, 10, and 60 min are reported in Table 7.13.

It can be seen that the UDSTW heuristic drastically improves the initial solutions in the first minute of computation. Different to the QCSP, further improvements are observed for all instance sets even after 10 min of runtime. However, these improvements are only marginal. Although the comparably high *ARE*s indicate further improvement potential, it can be concluded that the UDSTW heuristic delivers solutions of acceptable quality for the QCSPTW even if terminated after 1 min of runtime.

7.5 Summary

Within this chapter the problem of QC scheduling on the basis of container groups has been studied, as pioneered by Kim and Park (2004). The problem formulation considers crane scheduling in detail by incorporating crane interference constraints and by respecting travel time for crane movement. However, the QCSP model and also revised versions presented in later papers do not detect crane interference in every case. To derive a correct QCSP model, a set of new interference constraints has been formulated. A so-called UDS heuristic is used to solve the problem. It works on a reduced search space of unidirectional schedules. Computational tests demonstrate the power of the heuristic. It clearly outperforms all existing approaches to the QCSP with container groups, in terms of solution quality as well as in terms of computation times. It confirms that defining tasks by container groups leads to better solutions than task definitions on the basis of complete bays or bay areas. Furthermore, safety margins have a strong impact on the makespan of QC schedules and, therefore, need to be incorporated into practical QCSP formulations.

Moreover, a variant of the QCSP that respects time windows for cranes, referred to as the QCSPTW, has been formulated. The UDS heuristic has been adapted to solve this problem. The resulting UDSTW heuristic solves with difficulty medium-sized and large-sized QCSPTW instance within a runtime limit of 60 min. Nevertheless, the heuristic delivers solutions of good quality after a runtime of 1 min.

Summarizing this study, rich formulations for QC scheduling problems have been derived, which can be solved to good or even optimal solution quality within short runtimes by the proposed heuristics. These properties enable a functional integration of crane scheduling into the BACAP.

Chapter 8
Integration of Quay Crane Scheduling into the BACAP

Having investigated the BACAP, the QCSP, and the QCSPTW in detail in previous chapters, it is now turned to an integration of these problems. This study completes the realization of the integration concept that has been designed in Chap. 5. Section 8.1 outlines the functional integrations of crane scheduling into the berth planning phase in the context of practical application. Sections 8.2 and 8.3 describe the integration of QCSP and QCSPTW into the BACAP, respectively. Computational tests follow in Sect. 8.4. The study on integrated operations planning is concluded in Sect. 8.5.

8.1 Idea and Outline

In the BACAP model presented in Sect. 6.1 the assignment of crane capacity to vessels has been integrated into berth planning. An additional integration of QC scheduling issues is motivated by the following observation. The assignment of appropriate crane capacity depends on the realizable productivity of the cranes at a vessel, or, from another point of view, on the productivity loss caused by crane interference. Since the productivity of cranes is unknown in advance, a vessel might receive (i) insufficient crane capacity or (ii) superfluous crane capacity. In case (i), the handling time of a vessel needs to be extended, once the berth plan is executed. If the extension requires a delay in the berthing of other vessels, the berth plan is infeasible. In case (ii), a vessel can be served faster than the planned handling time, but the resulting improvement potential is not identified during the berth planning. In order to assign appropriate crane capacity to vessels it is necessary to provide a good estimate on crane productivity to the berth planning process.

In Sect. 6.1.2 an interference exponent has been introduced to model decreasing marginal productivity of cranes that simultaneously serve a vessel. This estimate is closer to reality than the one of Park and Kim (2003) who assume that the crane productivity is directly proportional to the number of assigned cranes. The computational Test 6.7 has verified the strong impact of the interference exponent on

F. Meisel, *Seaside Operations Planning in Container Terminals*,
Contributions to Management Science,
DOI: 10.1007/978-3-7908-2191-8_8,

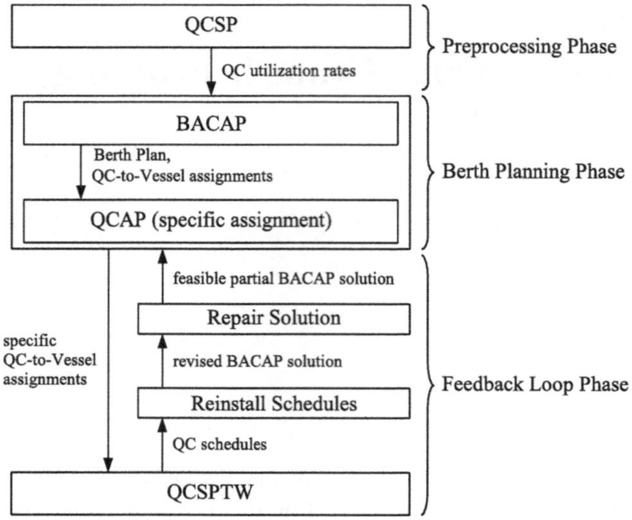

Fig. 8.1 Integration concept with an outline of the feedback loop steps

the BACAP solution quality, underlining the relevance of precise crane productivity estimation for the berth planning phase. Unfortunately, this kind of productivity estimation still shows a weakness. The interference exponent approximates an average crane productivity for the vessels. The productivity of cranes, however, differs among vessels. Even for vessels of same size and same number of containers to transship, the crane productivity is usually unequal if the vessels differ in the distribution of workload over the bays. Obviously, assigning appropriate crane capacities within the BACAP requires knowing vessel individual crane productivities. This chapter shows how a productivity estimate is obtained by solving a QCSP prior to the BACAP, see Fig. 8.1. This *preprocessing phase* provides vessel individual *QC utilization rates* which support the assignment of appropriate crane capacity to the vessels.

In the berth planning phase, QC-to-Vessel assignments are generated using the QC utilization rates. Unfortunately, even the vessel individual crane utilization rates provide merely a productivity estimate. To ensure that vessels receive crane assignments that actually allow for complete service, the QCSPTW is integrated into the berth planning phase using a *feedback loop*, see Fig. 8.1. Here, the QC-to-Vessel assignments derived in the berth planning phase serve as an input. They determine the availability of cranes for the vessels in the QCSPTW. The derived QC schedules indicate in turn, whether the QC-to-Vessel assignments represent sufficient crane capacity for the service of vessels. If assigned capacities are proven to be insufficient, crane assignment decisions have to be revised. This is done by reinstalling crane schedules in terms of matched QC-to-Vessel assignments. If the resulting berth plan is feasible, the feedback loop terminates and the planning process ends. Otherwise, if the reinstallation effects infeasibilities in a solution, a repair process restarts the berth planning phase and thus, completes the feedback loop.

Note that the preprocessing and the feedback loop serve different purposes. The preprocessing provides more precise input to the berth planning phase while the feedback loop ensures feasibility of solutions and completes the output of the seaside operations planning process by generating crane schedules.

Preprocessing and feedback loop are ways of functional integration, i.e., they are based on an agenda for jointly solving the optimization problems, see Sect. 5.1. Alternatively, a deep integration of BACAP and QCSP (QCSPTW) might be considered, but it fails because of the computational complexity. It is obvious that the resulting monolithic model cannot be solved by standard solvers such as CPLEX because already the BACAP model is much too complex to be solved within a reasonable runtime. Even a heuristic solution of the monolithic model, e.g., by incorporating a crane scheduling algorithm into the crane assignment procedure of the SWO algorithm, is computationally prohibitive. For instance, using SWO for planning the service of 40 vessels requires generating approximately 10^8 QC-to-Vessel assignments. Even a very fast heuristic like UDS cannot provide crane schedules for all these assignments within an acceptable computation time. As described in Sect. 5.3, deep integration fails also if stowage plans are not available for all vessels, whereas functional integrations are still applicable because they can be selectively bypassed for single vessels. For reasons of clarity, however, the following descriptions of functional integrations are aligned to a situation where stowage plans are available for all vessels.

8.2 Preprocessing Phase

Crane productivity can be estimated in two ways, see Fig. 8.2. A coarse crane productivity estimate stems from the usage of an interference exponent as used in the previous BACAP study. Alternatively, crane productivity can be estimated through utilization rates of cranes assigned to a vessel. In the following, it is described how to derive vessel individual crane utilization rates from solving a QCSP. Afterwards, the usage of utilization rates for crane productivity estimation and for making crane capacity assignment decisions is explained.

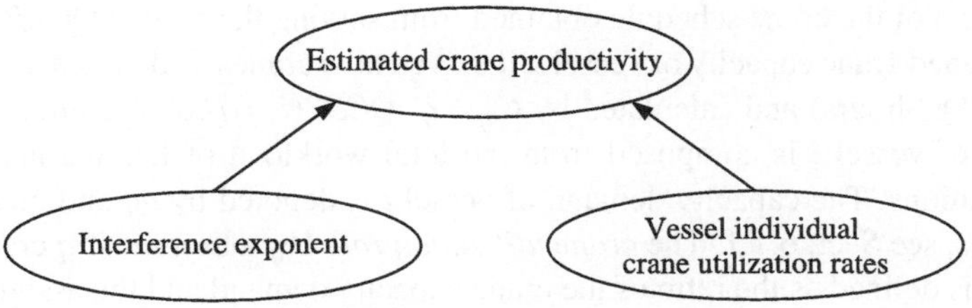

Fig. 8.2 Alternatives for an estimation of crane productivity

8.2.1 Deriving Crane Utilization Rates

An elegant way to use preprocessed crane schedules for improved berth planning and crane assignment is provided in Liu et al. (2006). The approach considers time-invariant QC-to-Vessel assignments, i.e., the number of cranes assigned to a vessel cannot change during the service process. Hence, the makespan of an optimal QCSP solution for the assigned number of cranes serves as the vessel's handling time in the berth planning process. In the BACAP, the number of cranes to assign to vessel i must be taken from the range $R_i = [r_i^{\min}, r_i^{\max}]$, where $r_i^{\min}$ and $r_i^{\max}$ denote the minimum and maximum number of assignable cranes. Solving the respective $|R_i|$ QCSP instances for each vessel of a BACAP instance leads to a set of potentially relevant handling times. For example, if a vessel can be served by 2, 3, or 4 cranes, three QCSP instances are preprocessed and the resulting three values for the makespan are provided to the berth planning phase. Depending on whether 2, 3, or 4 QCs are actually assigned to the vessel, the corresponding makespan value is selected as the vessel handling time in the berth plan.

It has been demonstrated in Sect. 6.3 that berth plans of better quality are obtained if variable-in-time crane assignments are allowed. Unfortunately, under a variable-in-time assignment, the handling time of a vessel can differ from every value generated by solving QCSP instances in the preprocessing. This is because a time-invariant crane assignment is assumed in the QCSP model. Circumventing this problem by preprocessing crane schedules for all possible variable-in-time QC-to-Vessel assignments is computationally prohibitive because the number of such assignments grows exponentially with the crane capacity demand of a vessel and the number of assignable cranes.

Consideration of variable-in-time QC-to-Vessel assignments within the berth planning phase necessitates that the preprocessing phase delivers crane utilization rates but no crane schedule makespan values. If, for example, the crane utilization rates for serving a vessel by 2, 3, or 4 cranes simultaneously are known, it can be estimated whether a variable assignment (e.g., assigning three cranes for 5 hours and four cranes for two more hours) represents sufficient crane capacity to complete the service within the projected service interval.

Vessel individual crane utilization rates are calculated as follows. The *crane capacity assigned* to a vessel is defined as the product of the number of assigned cranes and the handling time of the vessel. The handling time of vessel i, if served by q QCs, is denoted by $QCSP(i,q)$. It corresponds to the makespan (measured in minutes) of the crane schedule obtained from solving the related QCSP instance. The assigned crane capacity of vessel i, if served by q cranes, is denoted as C_{iq} (measured in QC-hours) and calculated by $C_{iq} = q \cdot QCSP(i,q)/60$. The *crane capacity demand* of vessel i is composed from the total workload of loading and unloading operations. The capacity demand of vessel i is denoted by m_i and measured in QC-hours, see Sect. 6.1.1. The *crane utilization rate* U_{iq}, observed if q cranes serve vessel i, is defined as the ratio of the crane capacity demand and the assigned crane capacity, i.e.,

$$U_{iq} = \frac{m_i}{C_{iq}} = \frac{m_i \cdot 60}{q \cdot QCSP(i,q)}. \tag{8.1}$$

For $U_{iq}=1$, the assigned crane capacity is fully exploited by the vessel service process. Since QCs are a bottleneck resource in CTs, a high crane utilization is desirable. Nevertheless, for reasons of crane interference and crane movement time, the crane capacity can usually not be fully exploited even under a perfect planning. Therefore, $U_{iq}<1$ is usually observed for the cranes.

Summarizing, the preprocessing phase contains the following two steps:

1. Solve a corresponding QCSP instance for each $i \in V, q \in R_i$
2. Calculate U_{iq} for each $i \in V, q \in R_i$ from the makespan obtained in the first step

A significant computational effort is only caused in the first step. For example, for a BACAP instance with 40 vessels where each vessel can be served by three different numbers of cranes, 120 QCSP instances need to be solved. Fortunately, a powerful QCSP solution method such as the presented UDS heuristic solves the majority of instances in negligible time. For this reason, the computational effort caused by the preprocessing is assumed to be acceptable in a practical application.

Example 8.1: Calculation of crane utilization rates

Consider vessel i with six bays and a workload of four container groups. The total processing time of the container groups is 300 min, i.e., the crane capacity demand is $m_i=5$ QC-hours. Assume that the vessel can be served by two or three QCs simultaneously. Solving the QCSP instances for $q=2$ and $q=3$ cranes yields the two schedules shown in Fig. 8.3. The corresponding makespan values are $QCSP(i,2)=192$ min and $QCSP(i,3)=110$ min, respectively. Using (8.1) the vessel individual crane utilization rates $U_{i,2}=(5\times 60)/(2\times 192)=0.78$ and $U_{i,3}=0.91$ are

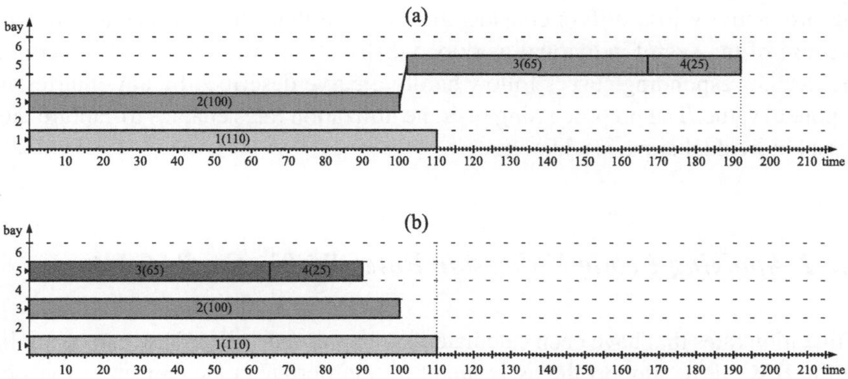

Fig. 8.3 Crane schedules for $q=2$ (**a**) and $q=3$ (**b**) QCs

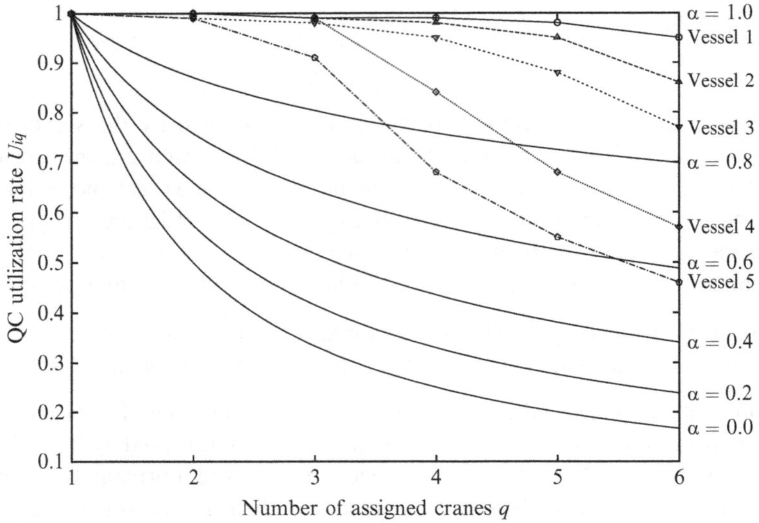

Fig. 8.4 Example crane utilization rates

calculated for $q = 2$ and $q = 3$, respectively. Surprisingly, the utilization rate is higher for $q = 3$ cranes than for $q = 2$ cranes. The example shows that the assignment of an additional crane can actually reduce productivity losses.

The consideration of vessel individual crane utilization rates is strongly motivated by the presumption that crane productivity loss depends on the particular distribution of workload over the bays of a vessel. To gain insight, five vessels of identical size of 25 bays are considered that have been generated for the computational tests at the end of this chapter. Fig. 8.4 shows crane utilization rates for the vessels if served by one to six QCs. For comparison, utilization rates are also computed for interference exponent values $\alpha = 0.0$ to $\alpha = 1.0$ (in steps of 0.2) by $U_{iq} = q^{\alpha-1}$. As can be seen in the figure, Vessels 1, 2, 3, and 4 can utilize up to three cranes without a significant productivity loss. Although the vessels are of same size, the productivity loss differs considerably if more than three cranes are assigned because of the vessel individual workload distribution over the bays. Furthermore, the five corresponding curves follow hardly a curve described by any interference exponent value. The preprocessing of crane utilization rates enables to capture these vessel individual crane productivities.

8.2.2 Applying Crane Utilization Rates Within the BACAP

Utilization rates that have been calculated in the preprocessing phase can be applied in the BACAP to support the assignment of appropriate crane capacity to vessels. If, however, a vessel receives a variable-in-time QC-to-Vessel assignment, it is not guaranteed that cranes meet the preprocessed crane utilization rates. Hence, crane utilization rates serve the purpose of a crane productivity estimation. More precisely, if q cranes are assigned to vessel i for 1 hour the expected effective crane capacity is $q \cdot U_{iq}$ QC-hours.

Two modifications are necessary to apply crane utilization rates in the presented BACAP model and the corresponding solution methods. These modifications affect those terms and constraints that contain the interference exponent. First, the minimum handling time needed to serve vessel i, as defined by (6.1), is redefined by

$$d_i^{\min} = \left\lceil \frac{(1+\beta\Delta b_i)m_i}{r_i^{\max} U_{i,r_i^{\max}}} \right\rceil. \tag{8.2}$$

Second, Constraint (6.3) of the BACAP model, used to decide on the crane capacities to assign to vessels, is reformulated as

$$\sum_{t \in T} \sum_{q \in R_i} (qU_{iq} r_{itq}) \geq (1+\beta\Delta b_i)m_i \quad \forall i \in V. \tag{8.3}$$

It can be seen that the modified BACAP model requires no additional variables or constraints and remains linear. The presented BACAP heuristics need no modification because the procedure *QC_Assignment*, used within the construction heuristic of Sect. 6.2.1, respects (8.2) and Constraints (8.3). It becomes obvious that the functional integration of the QCSP into the berth planning phase is based on a specification of input data rather than on advanced modeling techniques.

Example 8.2: Application of crane utilization rates (continued Example 8.1)

An example is given to illustrate the usage of preprocessed QC utilization rates for estimating the effective crane capacity of a variable-in-time QC-to-Vessel assignment. Consider again the vessel of Example 8.1. As shown there, utilization rates $U_{i,2} = 0.78$ and $U_{i,3} = 0.91$ are provided to the BACAP by the preprocessing. Assume that solving the BACAP determines a berthing time s_i and the berthing position $b_i = b_i^0$, i.e., the vessel is berthed at its desired berthing position. Assume furthermore that two QCs are assigned to the vessel for the first two handling periods and three QCs are assigned in a third handling period. All decisions taken from the solution of the BACAP are illustrated in Fig. 8.5. Recall that the assignment of QCs is controlled in the BACAP model by a binary variable r_{itq}, set to 1, if and only if exactly q QCs are assigned to vessel i at time t. Hence, for the vessel in this example, the crane assignment variables $r_{i,s_i,2}$, $r_{i,s_i+1,2}$, and $r_{i,s_i+2,3}$ are set to 1. With these values, Constraint (8.3) yields

$$\begin{aligned} & \sum_{t \in T} \sum_{q \in R_i} (qU_{iq} r_{itq}) \overset{!}{\geq} (1+\beta\Delta b_i)m_i, \\ \Rightarrow\ & 2 \times 0.78 + 2 \times 0.78 + 3 \times 0.91 \overset{!}{\geq} (1+\beta \times 0)5, \\ \Rightarrow\ & 5.85 \overset{!}{\geq} 5. \end{aligned}$$

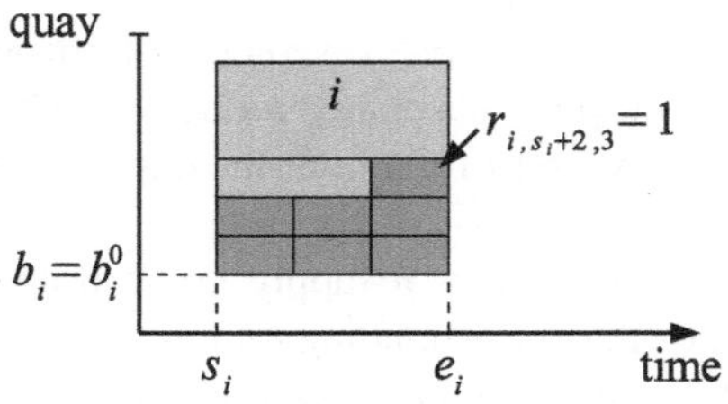

Fig. 8.5 A variable-in-time QC-to-Vessel assignment

Since the last inequality holds, the variable-in-time QC-to-Vessel assignment is expected to represent sufficient crane capacity for a complete service of the vessel.

8.3 Feedback Loop Phase

The outcome of the berth planning phase yields a berthing time, a berthing position, and a crane assignment for each vessel. The execution of the preprocessing is expected to improve the assignment of crane capacity to vessels. Unfortunately, there is no guarantee that vessels receive sufficient crane capacity because the assignment of cranes relies on productivity estimates. The aim of the feedback loop phase is to ensure that vessels receive crane assignments that allow for complete service.

As shown in Fig. 8.1, the feedback loop consists of three steps:

1. Generate QC schedules for a given BACAP solution by solving a QCSPTW
2. Reinstall the obtained QC schedules into the BACAP solution by matching the QC-to-Vessel assignments with the schedules
3. Repair infeasibilities by restarting the berth planning phase for the vessels that induce infeasibilities

The restart of the berth planning phase in the third step completes one round of the feedback loop. The loop terminates after the second step, if the reinstallation of crane schedules did not effect infeasibility of a solution. The three steps are described in detail below.

8.3.1 Postprocessing of a QCSPTW

In the postprocessing step of the feedback loop, crane schedules are generated for the vessels with respect to the crane assignments produced in the berth planning phase. Generally, the postprocessing comprises to solve a QCSP instance for each vessel which received a time-invariant QC-to-Vessel assignment and to solve a QCSPTW instance for each vessel which received a variable-in-time QC-to-Vessel assignment. For reasons of clarity, the following descriptions are aligned to a postprocessing which addresses the QCSPTW. The postprocessing step contains the following substeps:

1. Derive a QCSPTW instance for each vessel $i \in V$ from its QC-to-Vessel assignment given in the BACAP solution
2. Apply the UDSTW heuristic to the instances obtained in the first step

In the first substep, time windows for cranes are derived from a QC-to-Vessel assignment as described in Sect. 7.3.1. The productivity loss caused by berthing vessels apart from their desired berthing position is taken into account by increasing the processing times of loading and unloading tasks in the corresponding QCSPTW instance. The increase is calculated from the berth deviation factor β, see Sect. 6.1.2. In the second substep, the UDSTW heuristic is applied with a runtime limit according to the maximum acceptable response time. The outcome of the postprocessing step describes a QC schedule for each vessel.

Example 8.3: Postprocessing (continued Example 8.2)

Recall the vessel berthing at its desired berthing position with two cranes assigned for two periods and three cranes assigned in a third period. As shown in Example 8.2, the assigned capacity seems to be sufficient for the vessel's service. Assume in particular, that the assigned cranes are QCs 1, 2, and 3. Figure 8.6 shows the QC-to-Vessel assignment and the corresponding crane schedule generated in the postprocessing step. The derived schedule is a unidirectional one with a downward movement of the cranes. It has been found by the UDSTW heuristic using the problem data transformation described in Sect. 7.2.1. The schedule confirms that the vessel can be served within the handling time of 3 h as projected by the QC-to-Vessel assignment. The schedule further reveals that QC 1 is not busy within the third hour of service, and, as a consequence, only two cranes are needed for the service.

An alternative variable-in-time QC-to-Vessel assignment and the corresponding crane schedule are shown in Fig. 8.7. Here, QC 3 is assigned to the vessel in the first handling period instead of the third handling period. The QC-to-Vessel assignment is again expected to provide sufficient crane capacity because Constraint (8.3) holds again. Since tasks 1, 2, and 3 show comparably long processing times and task 4 cannot be started before its predecessor task is finished, the three cranes cannot

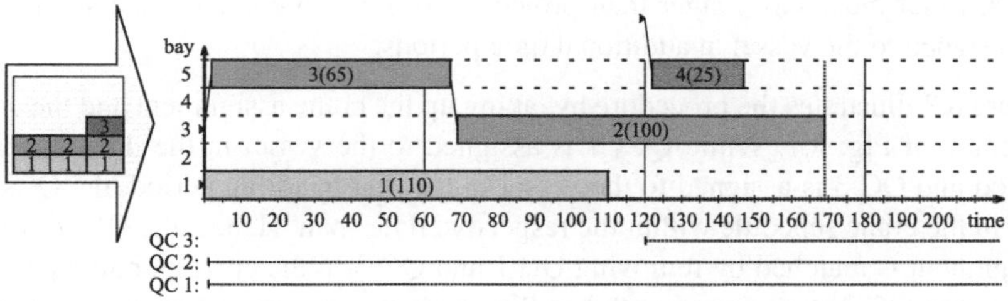

Fig. 8.6 A QC-to-Vessel assignment identified as sufficient by a corresponding QC schedule

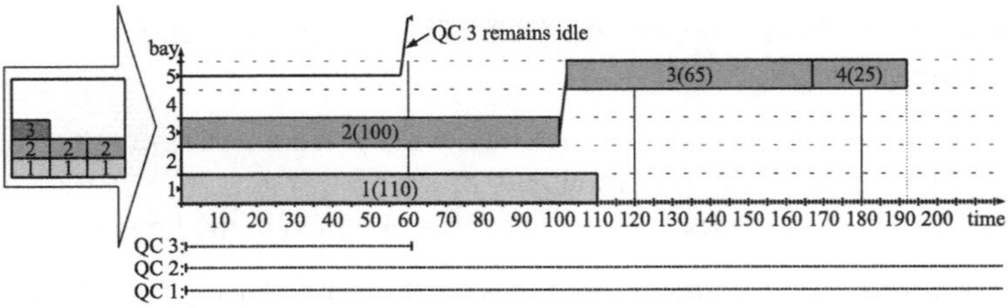

Fig. 8.7 A QC-to-Vessel assignment identified as insufficient by a corresponding QC schedule

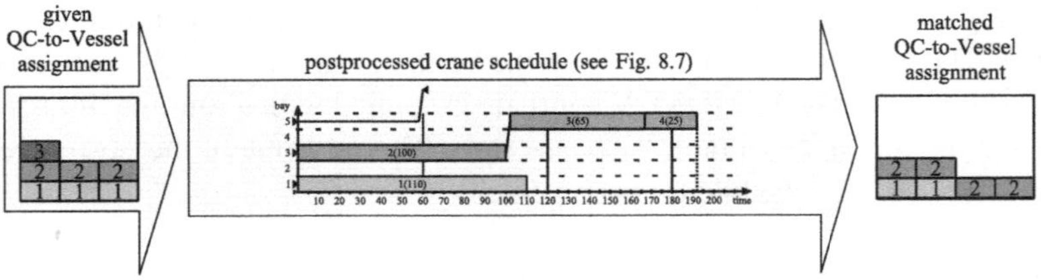

Fig. 8.8 Reinstalling crane schedules by matching QC-to-Vessel assignments

operate simultaneously at the vessel. Consequently, QC 3 is removed from the vessel at time 60 in the crane schedule without having processed any task. QC 2 must stay at the vessel for a fourth hour in order to complete the service. This demonstrates that not every QC-to-Vessel assignment, which has passed the capacity estimation check, is actually sufficient for a complete service of a vessel.

8.3.2 Reinstalling Quay Crane Schedules

If a postprocessed QC schedule does not fit the projected QC-to-Vessel assignment, the crane capacity assigned to the vessel is proven to be inappropriate for the service. The inconsistencies between the projected QC-to-Vessel assignment and the QC schedule need to be resolved. The following procedure is applied to match a crane assignment according to the utilization of cranes in the schedule:

- QCs that are idle in the crane schedule throughout one or more hours are removed from the current QC-to-Vessel assignment in the corresponding periods.
- QCs that must stay longer than projected in order to complete the service are assigned to the vessel in additional time periods.

Figure 8.8 illustrates the procedure by taking up the crane assignment and the crane schedule of Fig. 8.7. While QCs 1 is assigned to the vessel in the third handling period and QC 3 is assigned to the vessel in the first handling period, the QCs are idle in the crane schedule within the respective time span. Hence, the QC-to-Vessel assignment is matched by removing QC 1 and QC 3 in the corresponding periods. Moreover, QC 2 stays for a fourth handling hour in the crane schedule to complete the vessel's service and thus, the crane is also assigned for a fourth period in the matched QC-to-Vessel assignment.

Recall that a minimum number of cranes $r_i^{\min}$ may be contracted between the operator of vessel i and the CT management. However, a QC schedule obtained in the postprocessing step may reveal that this number of QCs cannot be utilized within every period. For example, in Fig. 8.8 only a single crane is busy in the third and the fourth period, although $r_i^{\min}=2$. In such a situation, vessel operators are assumed to agree removing idle cranes from their vessels because the service quality is not affected. Since QC operational cost can be saved this way, it is economically sensible to ignore strict contractual agreements in the BACAP model if not all cranes can be utilized.

Replacing the QC-to-Vessel assignments in the original BACAP solution by matched crane assignments yields a new BACAP solution, referred to as a *revised BACAP solution* in the following. Obviously, berthing times and berthing positions of vessels are not affected by this operation. The only changes possibly taking place are:

- The handling times of vessels change, causing earlier or delayed departures.
- The QC utilization changes within the periods.

To assess the differences between the original and the revised BACAP solution, six measures are introduced, denoted as M_1 to M_6. The measures M_1 and M_2 assess two types of infeasibilities that can be observed in a revised BACAP solution, namely overlapping rectangles in the space–time representation caused by delayed departures of vessels (M_1) and overshooting of the available QC capacity caused by matched QC-to-Vessel assignments (M_2). M_3 and M_4 assess changes in the service quality of a solution in terms of cost and departure times. Finally, measures M_5 and M_6 quantify changes in the QC utilization over the entire planning horizon and on a periodical basis.

To formalize these measures it is supposed that the original BACAP solution is completely represented by the decision variables $e_i, u_i, \Delta ETA_i$, and ΔEFT_i as introduced in Sect. 6.1. For a compact description, the number of cranes assigned to vessel i at period t is represented by $q_{it} = \sum_{q \in R_i} (q \cdot r_{itq})$. From this formula it can be seen that $q_{it}=0$ holds, if and only if $r_{itq}=0$ holds for all $q \in R_i$. The revised BACAP solution is represented by corresponding variables e_i', u_i', $\Delta ETA_i'$, $\Delta EFT_i'$, and q_{it}'.

The Feasibility Measures M_1 and M_2

M_1 – Overlapping in the space–time representation: The measure M_1 yields the total number of segments in the space–time representation of the revised BACAP solution that are occupied by more than one vessel. Let o_{lt} be 1 if and only if the segment $(l,t), l \in \{0,1,\ldots,L-1\}, t \in T$ is occupied by more than one vessel. M_1 is calculated by

$$M_1 = \sum_{l=0}^{L-1} \sum_{t \in T} o_{lt}. \tag{8.4}$$

M_2 – Overshooting of QC capacity: The measure M_2 yields the total number of QC-hours that overshoot the available QC capacity within the planning horizon. With Q as the number of QCs available at the terminal, M_2 is calculated by

$$M_2 = \sum_{t \in T} \max \left\{ 0, \sum_{i \in V} q'_{it} - Q \right\}. \tag{8.5}$$

The Service Quality Measures M_3 and M_4

M_3 – Change in service quality cost: The measure M_3 yields the change in service quality cost over all vessels. It is calculated by

$$M_3 = \sum_{i \in V} \left(c_i^1 \Delta ETA'_i + c_i^2 \Delta EFT'_i + c_i^3 u'_i \right) - \sum_{i \in V} \left(c_i^1 \Delta ETA_i + c_i^2 \Delta EFT_i + c_i^3 u_i \right). \tag{8.6}$$

M_4 – Absolute change in departure times: Early and delayed departures of vessels may cancel out service quality cost changes in M_3. Therefore, measure M_4 yields the total absolute change in the departure times of vessels. M_4 is calculated by

$$M_4 = \sum_{i \in V} \left| e'_i - e_i \right|. \tag{8.7}$$

The QC Utilization Measures M_5 and M_6

M_5 – Change in total utilized QC-hours: M_5 yields the change in total utilized QC-hours over the planning horizon. M_5 is calculated by

$$M_5 = \sum_{t \in T} \sum_{i \in V} q'_{it} - \sum_{t \in T} \sum_{i \in V} q_{it}. \tag{8.8}$$

M_6 – Absolute change in utilized QC-hours on a periodical basis: A canceling out effect is observed for measure M_5 if some periods show an increase in crane capacity utilization while other periods show a decrease in crane capacity utilization. Therefore, M_6 sums up the absolute change in utilized QC-hours within the 1-hour periods of the planning horizon. M_6 is calculated by

$$M_6 = \sum_{t \in T} \left| \sum_{i \in V} q'_{it} - \sum_{i \in V} q_{it} \right|. \tag{8.9}$$

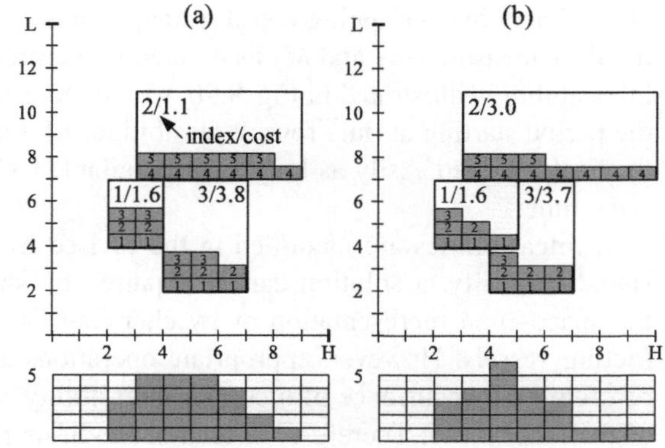

Fig. 8.9 Original BACAP solution (**a**) and revised solution with matched QC-to-Vessel assignments (**b**)

Example 8.4: Assessing the revisions of a BACAP solution

The BACAP solution shown in Fig. 6.7 on page 67 is taken up and completed by a specific QC assignment, see Fig. 8.9a. The solution has been generated using an interference exponent $\alpha = 0.9$ for estimating crane productivity. Assume that QC schedules have been generated for the three vessels in the postprocessing step that lead to the revised BACAP solution shown in Fig. 8.9b. The differences between the original solution and the revised solution are assessed by measures M_1 to M_6. Overlapping is observed on $M_1 = 3$ space–time segments. It is caused by Vessels 1 and 3 in the period starting at time four. The only overshoot of QC capacity is observed in the same period and accounts for $M_2 = 1$ QC-hour. As measured by $M_3 = 2$, the total change in service quality cost is 2,000 USD. The increase of cost is caused by the delayed departure of Vessel 2. Other changes in cost observed for the revised BACAP solution are induced by changes in the number of utilized QCs. The absolute deviation in departure times of vessels add up to $M_4 = 2$ h, where Vessels 1 and 2 show a delay of one hour each. The change in total utilized QC-hours is $M_5 = -2$ QC-hours resulting from 25 QC-hours utilized in the original berth plan and 23 QC-hours utilized in the revised berth plan. On a periodical basis, the change in utilized QC-hours is $M_6 = 6$ QC-hours, composed from the changes of one QC-hour within each of the periods starting at time 3, 4, 5, 6, 7, and 9.

8.3.3 Repairing Infeasible BACAP Solutions

A revised BACAP solution is detected to be infeasible by at least one of the indicators:

1. Overlapping of vessel rectangles in the space–time representation
2. Overshooting of the available QC capacity
3. Assigning specific cranes to multiple vessels in a same period

Occurrences of overlapping vessel rectangles and crane capacity overshots are quantified by measures M_1 and M_2 as defined in the previous section. The third type of infeasibility is illustrated in Fig. 8.9b, where QC 1 is assigned to Vessels 1 and 3 in the period starting at time four. Infeasibilities of this type are not assessed because they are resolved easily as long as the available QC capacity is not overshot at the same time.

If infeasibilities are identified in the revised BACAP solution a repair process starts. Basically, a solution can be repaired by shifting vessel rectangles within the space–time representation or by changing QC-to-Vessel assignments of conflicting vessels. However, appropriate operations are difficult to develop because resolving the occurrence of a certain infeasibility may cause infeasibilities of one of the other types. Therefore, instead of modifying an infeasible solution, one can repeat the solution process of the BACAP under modified conditions. One possible option is to vary input data, e.g., by increasing or by decreasing the crane capacity demand m_i of those vessels for which an insufficient or superfluous QC-to-Vessel assignment is observed. Unfortunately, the effect on a BACAP solution is hardly predictable and the danger of cycling between multiple infeasible solutions is high. To evade this difficulty, a repair procedure is proposed that is based on fixing a feasible partial solution and restarting the BACAP solution process for the non-fixed part of the berth plan. This is repeated until a feasible BACAP solution is obtained. Thereby, the fixed partial solution is steadily extended which avoids cycling between infeasible solutions.

The repair procedure works as follows. It first identifies the types of infeasibility of the given BACAP solution. Of course, if no infeasibility exists at all, the solution requires no repair and the feedback loop is terminated. If the solution is only infeasible because specific QCs are assigned to multiple vessels in a same period, the conflict is easily resolved by calling the dynamic programming method of Park and Kim (2003), see Sect. 6.2.4. This method decides on the specific cranes that make up the assigned QCs of a vessel. Applying the method to the revised BACAP solution will lead to a new specific crane assignment where no cranes are assigned to multiple vessels in a same period. Afterwards, in a new iteration of the feedback loop, the postprocessing step is repeated to generate new crane schedules for the revised assignments.

Overlapping vessel rectangles in the space–time representation and overshooting of QC capacity are repaired as follows. The repair procedure identifies the earliest time period where the revised BACAP solution becomes infeasible. The beginning of this period is referred to as the *conflict time*. The partial berth plan defined by the subset of vessels that have a berthing time earlier than the conflict time is fixed. The non-fixed vessels are removed from the solution in order to resolve overlapping conflicts as well as overshooting of the QC capacity. Note that vessels with service being in progress at the conflict time have not been removed and may still cause capacity conflicts. These infeasibilities are resolved by iteratively removing cranes from vessels until the available crane capacity is reached. If vessels of different type are involved in such a conflict, feeder vessels receive priority against medium vessels and jumbo vessels in the removing process. The obtained feasible partial berth plan is extended by restarting the berth planning phase as outlined in Fig. 8.1. Here, the BACAP solution methods have been modified in order to schedule only non-fixed vessels. The scheduling respects the occupation of quay space and the utilization of QCs of the already fixed vessels. Then, specific QC-to-Vessel assignments are generated and the postprocessing is repeated. Computation time can be saved by repeating the postprocessing step only for those vessels that show a change in either the QC-to-Vessel assignment or in the berthing position. For all other vessels, the QC schedule built within the previous round is kept. The QC schedules are reinstalled in the BACAP solution, and the repair process is repeated if the solution is still infeasible. Since the conflict time moves steadily towards the planning horizon, a feasible berth plan is obtained in the end.

Example 8.5: Repairing an infeasible BACAP solution (continued Example 8.4)

The repair of the infeasible solution shown in Fig. 8.9b is illustrated. The solution is detected to be infeasible by all three types of infeasibility. To repair the solution, the conflict time is determined at time four, see Fig. 8.10a. At that time the first infeasibility occurs. The partial berth plan with Vessels 1 and 2 is fixed and Vessel 3 is removed from the solution. Rescheduling Vessel 3 by applying the SWO heuristic while keeping the fixed partial berth plan yields the revised solution shown in Fig. 8.10b. Since the repair process did not change the crane assignments for Vessels 1 and 2, the previously generated QC schedules are still usable. While the repair changed the berthing time of Vessel 3, its crane assignment and berthing position remain unchanged. For this reason, no new crane schedule must be generated for Vessel 3, too. Its QC-to-Vessel assignment is matched to the crane schedule, which leads to the feasible BACAP solution shown in Fig. 8.10c. Note that the repair is realized by delaying the service of Vessel 3, which leads to an increase in its service quality cost.

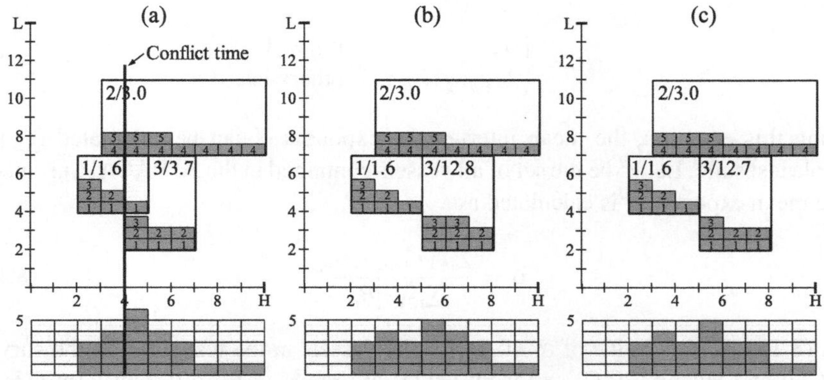

Fig. 8.10 Repairing an infeasible BACAP solution

8.4 Computational Study

The following tests assess the contributions of the preprocessing phase and the feedback loop phase to the seaside operations planning process:

Test 8.1: Comparison of QC productivity estimators
Test 8.2: Comparison of final solutions

Benchmark Instances

The BACAP instances generated for the computational study in Sect. 6.3 serve as a basis for the tests. Additionally, container group information needs to be given for the vessels contained in these instances. It is generated as follows: For every vessel of the feeder, medium, and jumbo class, a number of container groups is randomly drawn from $U[10,15]$, $U[15,35]$, and $U[35,50]$, respectively, where U expresses a uniform distribution of integer values in the specified interval. For the container groups, processing times are randomly drawn from $U[3,180]$ as proposed by Kim and Park (2004). The generation process has to ensure that the total processing time of the container groups of a vessel corresponds to the given crane capacity demand of the vessel. This is realized by a proportional scaling of the processing times. Afterwards, the container groups are randomly assigned to bays. The number of bays of a vessel i is approximated by $\lfloor l_i \cdot 10/12 \rfloor$, where l_i denotes the length of the vessel (given in segments of 10-m). It takes into account that a Forty-foot container bay is of about 12 m length. While randomly assigning container groups to bays, a bay can remain empty or receive one or more container groups. If more than one container group belongs to a bay, precedence constraints are randomly inserted between the groups to define a unique processing order.

In order to check to what extent the new container group distribution involves productivity loss by crane interference, individual crane utilization rates U_{iq} are determined for the vessels in the test instances. For vessel i, served by q cranes, an individual interference exponent α_{iq} is derived from U_{iq} as

$$\alpha_{iq} = \begin{cases} 1, & \text{if } q = 1, \\ \log_q(q \cdot U_{iq}), & \text{otherwise.} \end{cases} \tag{8.10}$$

Using this equation, the mean interference exponent $\overline{\alpha}$ can be calculated for the whole test suite. Let V' be the set of all vessels contained in the 30 BACAP instances. The mean exponent $\overline{\alpha}$ is calculated as

$$\overline{\alpha} = \frac{\sum_{i \in V'} \sum_{q \in R_i} \alpha_{iq}}{\sum_{i \in V'} |R_i|}. \tag{8.11}$$

Eq. (8.11) yields a value of $\overline{\alpha} = 0.94$ for the vessels in the test suite. In the survey of Chu and Huang (2002) exponent values are reported for different terminals of the port of Kaohsiung, ranging from 0.80 to 1.00 with a mean of 0.93. It can be concluded from this that the generated container group information appears quite realistic.

Test 8.1: Comparison of QC productivity estimators

Crane utilization rates derived from solving a QCSP in the preprocessing phase provide an interesting alternative against the crane productivity estimation based on an interference exponent. In order to compare both alternatives, the above described test suite is used. Each of the 30 instances is solved using every exponent value α from 0.80 to 1.00 (in steps of 0.02) and using the preprocessed crane utilization rates U_{iq}. The SWO heuristic is applied within the berth planning phase to solve the BACAP. The solutions of the berth planning phase have to be assessed in order to identify the best performing productivity estimator. This is realized by a partial execution of the feedback loop. Only the postprocessing step and the reinstallation step of the feedback loop phase are applied to the solutions of the berth planning phase, i.e., a first revision of BACAP solutions is performed. The differences between the original and the revised BACAP solutions are assessed by the measures M_1 to M_6. Obviously, the better a particular crane productivity estimator performs, the less differences will be observed between the original and the revised BACAP solutions.

Figure 8.11 shows the results for the feasibility measure M_1. Results are reported for each investigated exponent value α and for the preprocessed crane utilization rates in column *QCSP*. Results are presented for a particular crane productivity estimator in the following way. The horizontal line within the gray shaded box represents the average value measured for M_1 in the test suite. The box itself is dimensioned by the 10% and the 90% quantile to give an idea of the spread of the observed values. The vertical lines indicate the minimum and maximum observed value for the 30 test instances.

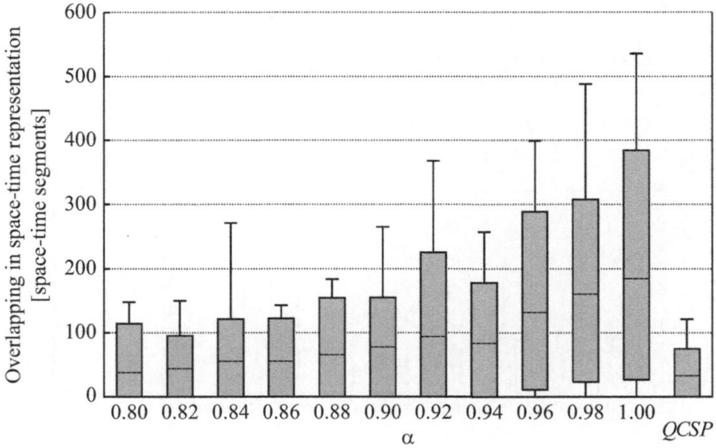

Fig. 8.11 Comparison of productivity estimators by measure M_1

The following results are observed for measure M_1. The lower the estimated crane productivity (small values of α), the less overlapping is observed in the space–time representation of the revised solution. Obviously, underestimating crane productivity increases the crane capacity assigned to vessels in the berth planning phase, which leads to long vessel handling times. Generating crane schedules in the postprocessing step identifies useless crane capacity. Consequently, matching the QC-to-Vessel assignments to the crane schedules shortens the projected handling times of vessels considerably without causing overlapping in the space–time representation. In contrast, if crane productivity is overestimated by a high value of α, the postprocessing effects increased handling times in the revised BACAP solution, which make overlapping more likely. If crane productivity is estimated using preprocessed crane utilization rates (column *QCSP*), BACAP solutions show on average 35 overlapping space–time segments. Notice that only interference exponents $\alpha \leq 0.82$ show similar values for M_1.

The number of QC-hours overshooting the available crane capacity at a terminal is measured by M_2. The observed results for this measure are provided in Fig. 8.12. Again, small values of α lead to few infeasibilities. The explanation is as for M_1. If crane productivity is underestimated, the crane capacity assigned to vessels will be reduced in the revised solution. Only few vessels require additional crane capacity, which makes it rather unlikely to overshoot the available capacity. Overshooting of 4 QC-hours is observed on average for a BACAP solution in column QCSP. Merely interference exponents smaller than 0.86 yield better average M_2 values.

Summarizing the results for measures M_1 and M_2, the preprocessing phase enables BACAP solutions which show only few infeasibilities. Small values of the interference exponent are apparently competitive with respect to both measures.

The results for the service quality measures M_3 and M_4 are shown in Figs. 8.13 and 14 respectively. M_3 measures the change in service quality cost. The average

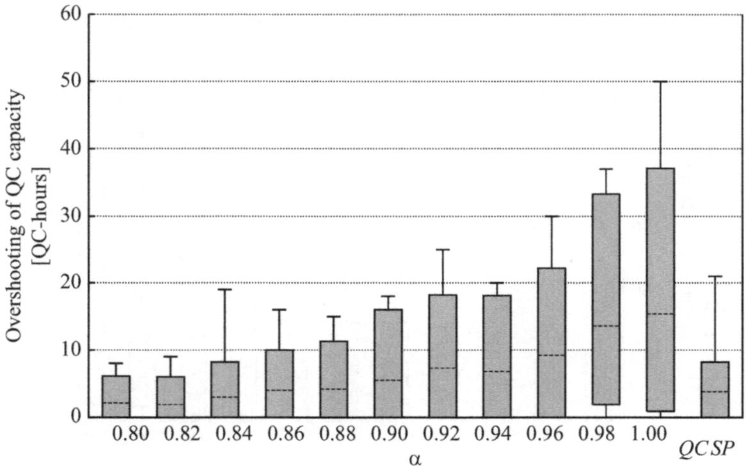

Fig. 8.12 Comparison of productivity estimators by measure M_2

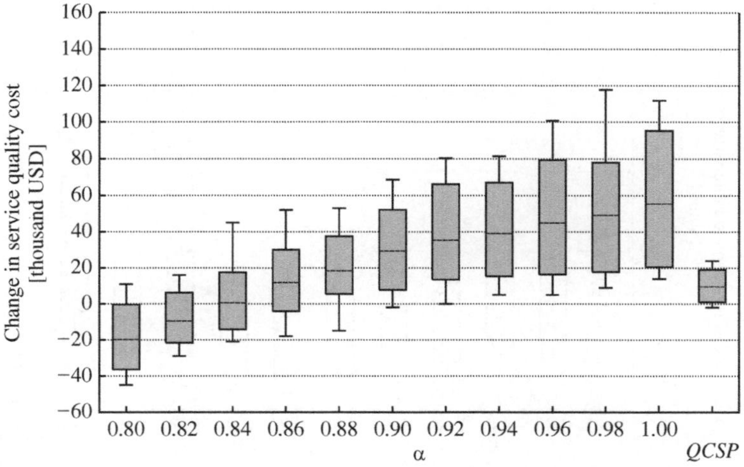

Fig. 8.13 Comparison of productivity estimators by measure M_3

M_3 value falls below zero for $\alpha \leq 0.82$. This means that a low crane productivity estimate leads to high service quality costs in the BACAP solutions, which are reduced afterwards in the revision. In contrast, overestimating QC productivity by a value of α close to one leads to low service quality costs in solutions. However, these cost increase in the revised solution because vessel handling times need to be increased there. Solutions with an average cost change close to zero are generated using $\alpha = 0.84$. If crane productivity is determined in a preprocessing phase, a slight increase in service quality cost is observed for revised BACAP solutions.

Service quality costs of vessels may cancel out in M_3 if the postprocessing causes some vessels to depart earlier, while other vessels depart later than projected in the original BACAP solution. Therefore, measure M_4 considers the absolute change in the departure times of the vessels. Considering results for the interference exponent, values of $0.88 \leq \alpha \leq 0.92$ lead to a common average M_4 value of about 20 h per instance, see Fig. 8.14. Larger changes in departure times as observed for $\alpha \leq 0.86$ ($\alpha \geq 0.94$) are based on considerable reductions (increases) of the vessel handling times in the postprocessing. In light of these results, the good performance of $\alpha = 0.84$ regarding measure M_3 is based to a certain extent on canceling-out effects. BACAP solutions that use preprocessed crane utilization rates show only little changes in vessel departure times. None of the interference exponent values delivers as good results.

Summarizing the measurement of service quality, the interference exponent is seldom competitive with preprocessed crane utilization rates. Regarding measure M_3, solutions using the preprocessed crane utilization rates show changes in service quality cost very close to zero. Merely an interference exponent of $\alpha = 0.84$ leads to better results. Regarding the change in vessel departure times (M_4), none of the investigated exponent values is competitive.

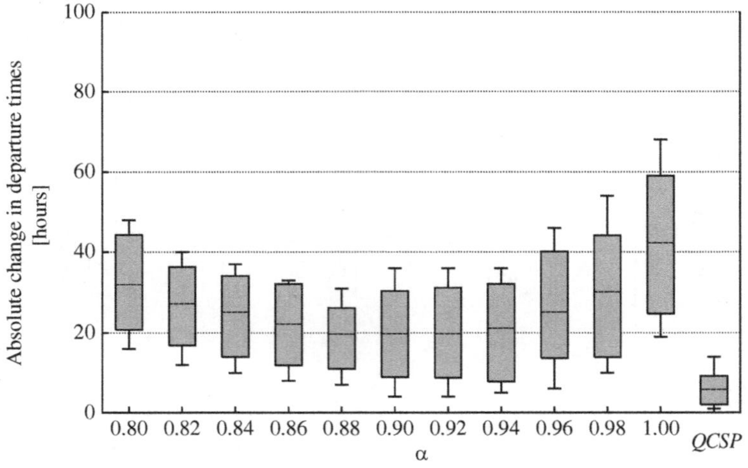

Fig. 8.14 Comparison of productivity estimators by measure M_4

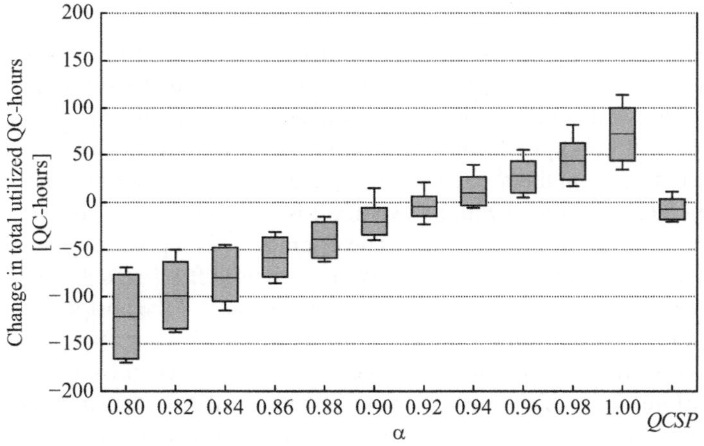

Fig. 8.15 Comparison of productivity estimators by measure M_5

The results for the QC utilization measures M_5 and M_6 are shown in Figs. 8.15 and 8.16. They are similar to the results observed for measures M_3 and M_4. The change in total utilized QC-hours, as measured by M_5, is negative (positive) for small (large) values of α because crane capacity is removed from (added to) vessels in the revised solutions. The best performing exponent values are $\alpha = 0.92$ and $\alpha = 0.94$. For these two values, the average change in utilized QC-hours is -5 QC-hours and $+10$ QC-hours per instance, respectively. The results show that a properly chosen interference exponent performs well to estimate the average crane productivity. The preprocessing phase enables solutions with an average change of -7 QC-hours per instance. With the only exception of $\alpha = 0.92$ and $\alpha = 0.94$, the interference exponent leads to considerably larger deviations of measure M_5 from zero.

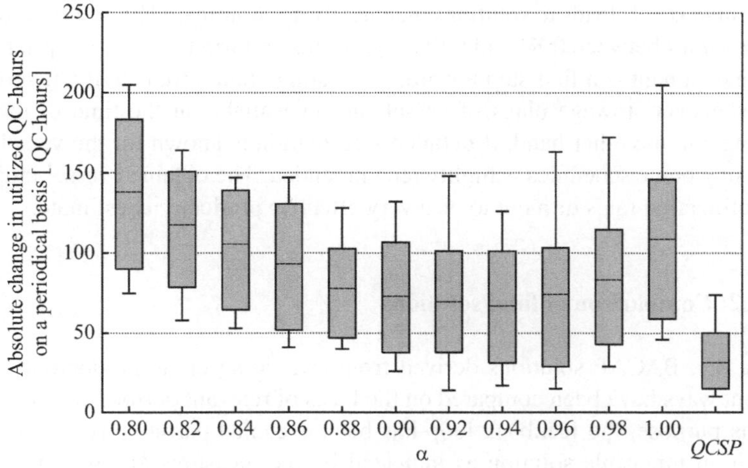

Fig. 8.16 Comparison of productivity estimators by measure M_6

As argued before, the total change in utilized QC-hours measured by M_5 may cancel out positive and negative changes in the crane utilization of different periods. Therefore, measure M_6 is used to assess the absolute change in the utilized QC-hours on a periodical basis. Fig. 8.16 shows a similar trend as Fig. 8.14. Again, the results reported in column QCSP dominate the results of all interference exponent values. This reveals that the good performance of $\alpha = 0.92$ and $\alpha = 0.94$ regarding measure M_5 stems from a canceling-out effect.

For the two QC utilization measures M_5 and M_6, it can be concluded that a moderate exponent value like $\alpha = 0.92$ is apparently competitive with the preprocessed utilization rates with respect to measure M_5. Regarding measure M_6, no interference exponent value enables results as good as a preprocessing does.

Summarizing the observations for all six criteria, BACAP solutions that use preprocessed utilization rates for the crane productivity estimation show a small number of infeasibilities (M_1, M_2) and only little changes in service quality (M_3, M_4) as well as in the crane utilization (M_5, M_6). Regarding measures M_4 and M_6, the solutions using preprocessed utilization rates dominate the results of all interference exponent settings. For the measures M_1, M_2, M_3, and M_5 there are α values which might be considered competitive. However, exponent values performing well for some of the measures perform extremely poor regarding the remaining measures. For example, comparing $\alpha = 0.82$ with the preprocessing shows that the exponent performs slightly better regarding M_2, similar regarding M_1 and M_3, but extremely poor regarding M_5. In general, small values of α lead to a waste of crane capacity and to bad service quality, whereas values of α near one cause infeasibilities at a high rate because vessels receive insufficient crane capacity. Nevertheless, also the mean interference exponent $\overline{\alpha} = 0.94$, as calculated above for the totality of vessels contained in the test suite, delivers satisfying results with respect to measure M_5 only. Ignoring productivity loss completely, as simulated by setting the interference exponent to $\alpha = 1$, leads to solutions that are always dominated by solutions derived from α values between 0.84 and 0.98. The results confirm that an appropriate interference exponent is a first step towards estimating crane productivity. It should be used whenever stowage plans of vessels are unavailable at the time of the berth planning. On the other hand, if detailed information is known for the vessels, preprocessing crane schedules is highly recommended. The obtained vessel individual crane utilization rates turn out to be a very effective productivity estimate.

Test 8.2: Comparison of final solutions

In Test 8.1, BACAP solutions derived from estimating crane productivity in two different ways have been compared on the basis of relevant performance measures. For this purpose, the feedback loop has been executed partially, typically ending up with an infeasible solution as indicated by the measures M_1 and M_2. A final feasible solution to the overall problem of seaside operations planning is obtained by continuing the feedback loop until all infeasibilities in a revised BACAP solution are eliminated.

Table 8.1 reports the service cost Z of final solutions for the 30 BACAP instances, where crane productivity is estimated using interference exponents $\alpha = 0.80$, $\alpha = 0.92$, $\alpha = 1.00$ and preprocessed crane schedules (column *QCSP*), respectively. The exponent value $\alpha = 0.80$ has been selected because it leads to solutions showing least infeasibilities in the previous test. The exponent value $\alpha = 0.92$ has been selected because it performed best in estimating average crane productivity for the vessels in the test suite. Furthermore, final solutions are reported where crane productivity loss has been completely ignored within the berth planning phase ($\alpha = 1.00$). For the solutions that base on the interference exponent, a relative error *RE* against the service cost Z in column *QCSP* is reported.

The results in Table 8.1 show that the best feasible solution to a problem is obtained in almost every case using preprocessed crane schedules. Only $\alpha = 0.92$ leads to better solutions for three of the small-sized instances (#2, 5, 7), one medium-sized instance (#16), and one large-sized instance (#26). The overall performance of each interference exponent value is revealed by the observed average relative error *ARE*. For $\alpha = 0.92$, solutions are obtained with service quality cost that are on average 11.58% above the cost of the solutions basing on preprocessed crane utilization rates. For $\alpha = 0.80$ and $\alpha = 1.00$ the average service cost of solutions are clearly larger. The *ARE* is about 25% for both exponent values. This shows that an inappropriate setting of the interference exponent ($\alpha = 0.80$) leads to solutions that are not better than solutions obtained by completely ignoring crane productivity loss ($\alpha = 1.00$).

Table 8.1 additionally reports the runtime of the feedback loop for the solutions using preprocessed crane schedules in column *time* in seconds. As one could expect, the runtime increases considerably from small-sized to large-sized instances. Obviously, in scenarios with high utilization of quay space and quay cranes, infeasibilities are more likely and require more loop iterations to be executed. Nevertheless,

Table 8.1 Comparison of final solutions

n	#	$\alpha = 0.80$		$\alpha = 0.92$		$\alpha = 1.00$		*QCSP*	
		Z	RE	Z	RE	Z	RE	Z	*time*
20	1	141.2	26.07	129.4	15.54	132.3	18.13	112.0	150
	2	63.2	0.48	61.1	−2.86	74.0	17.65	62.9	130
	3	99.5	23.45	88.5	9.80	133.1	65.14	80.6	282
	4	107.5	27.82	88.0	4.64	89.6	6.54	84.1	312
	5	87.5	12.04	76.1	−2.56	88.2	12.93	78.1	132
	6	78.3	6.53	77.5	5.44	85.2	15.92	73.5	132
	7	81.2	27.67	63.3	−0.47	74.5	17.14	63.6	177
	8	72.7	7.07	68.6	1.03	69.7	2.65	67.9	393
	9	81.3	15.48	72.0	2.27	82.8	17.61	70.4	248
	10	127.3	10.60	126.6	9.99	141.7	23.11	115.1	381
30	11	190.8	11.45	215.6	25.93	227.3	32.77	171.2	829
	12	144.4	58.33	106.6	16.89	120.1	31.69	91.2	391
	13	150.4	20.13	136.9	9.35	157.4	25.72	125.2	353
	14	146.7	12.59	142.0	8.98	151.7	16.42	130.3	514
	15	220.4	16.37	207.8	9.71	232.3	22.65	189.4	910
	16	137.9	11.12	122.1	−1.61	130.2	4.92	124.1	308
	17	151.3	21.04	144.4	15.52	161.7	29.36	125.0	499
	18	189.8	23.33	178.1	15.72	201.3	30.80	153.9	570
	19	256.8	31.56	208.9	7.02	232.7	19.21	195.2	543
	20	164.2	16.21	164.7	16.56	188.4	33.33	141.3	496
40	21	389.1	33.25	309.1	5.86	338.9	16.06	292.0	1,358
	22	247.9	38.11	259.8	44.74	271.3	51.14	179.5	524
	23	458.4	55.02	335.8	13.56	343.3	16.10	295.7	721
	24	516.4	46.79	428.0	21.66	446.3	26.86	351.8	1,602
	25	289.5	55.65	211.9	13.92	230.1	23.71	186.0	871
	26	354.2	13.56	307.8	−1.31	353.4	13.31	311.9	1,717
	27	301.9	15.41	282.0	7.80	312.4	19.42	261.6	1,907
	28	483.5	47.45	370.4	12.96	422.4	28.82	327.9	2,201
	29	330.7	31.96	331.8	32.40	388.2	54.91	250.6	799
	30	395.3	38.80	367.0	28.86	401.4	40.94	284.8	1,136
ARE(%)			25.18		11.58		24.50		

the observed runtimes of few minutes up to about half an hour per instance appear reasonably short. In particular the runtime limit applied for the UDS and UDSTW heuristics in the postprocessing step determines the computational effort of the feedback loop phase. The overall runtime to obtain a final BACAP solution can be adjusted in this way to fit the needs of a practical application. Summarizing this test, the feedback loop is effective in repairing BACAP solutions that turned out to be infeasible after the reinstallation of crane schedules.

8.5 Summary

Within this chapter, two functional integrations of crane scheduling into the berth planning and crane assignment process have been investigated. One integration results from preprocessing crane schedules in order to generate vessel individual crane utilization rates. These utilization rates serve as a crane productivity estimate in the berth planning phase. A computational test has compared this approach with a crane productivity estimation using an interference exponent. It was found that underestimating the value of the exponent leads to a waste of crane capacity, whereas overestimating the value of the exponent leads to insufficient assignment of crane capacity to the vessels. However, even moderate settings of the interference exponent, as suggested by empirical studies, are only competitive with respect to some of the relevant performance measures. Hence, it is concluded that the crane utilization rates provided by the preprocessing phase outperform the crane productivity estimation by an interference exponent.

A further functional integration is realized by a feedback loop, which iteratively revises berth plans on the basis of induced crane schedules. It ensures that each vessel receives sufficient crane capacity for its service. First, a postprocessing is used to generate QC schedules for the QC-to-Vessel assignments of a BACAP solution. These schedules are used to identify inappropriate crane capacity assignments. The necessary adaptations can lead to infeasibilities of a BACAP solution that are resolved in a repair process. The computational test of the feedback loop has shown that infeasibilities are effectively resolved. The best final solutions are obtained if crane productivity is estimated in the preprocessing phase.

Summarizing, estimating crane productivity by preprocessing crane schedules is worth the effort spent. The berth planning phase benefits from the specified crane productivity information. The feedback loop effectively ensures that vessels receive sufficient crane capacity. The crane schedules that are generated within the loop complete the outcome of the seaside operations planning process. Moreover, high quality final solutions to the overall problem of seaside operations planning are obtained from the joint application of the preprocessing phase and the feedback loop phase. Hence, both functional integrations support seaside operations planning.

Chapter 9
Conclusions

The thesis deals with seaside operations planning in seaport container terminals. The investigated planning problems are the Berth Allocation Problem (BAP), the Quay Crane Assignment Problem (QCAP), and the Quay Crane Scheduling Problem (QCSP). From solving these problems, berthing times, berthing positions, and crane capacity are assigned to vessels, and quay crane schedules are determined. The decisions are closely interrelated due to the strong dependence of vessel handling times on the assignment and scheduling of cranes. Nevertheless, the planning problems are predominantly considered isolated in scientific research. In practice they are solved in a sequential solution process, which is hardly able to respect the interrelations and thus, threatens a good utilization of terminal resources and a corresponding service quality. The aim of this research is to overcome these weaknesses by providing an integrated solution approach for seaside operations planning.

The first scientific contribution of the thesis to container terminal operations management is the provision of classification schemes for the various BAP, QCAP, and QCSP formulations presented in the literature. Using these schemes, all approaches can be classified and the essential differences are uncovered.

As a second scientific contribution, concepts have been identified for the integrated planning of berth allocation, crane assignment, and crane scheduling. A new concept has been designed, which comprises a deep integration of BAP and QCAP, leading to the Berth Allocation and Crane Assignment Problem (BACAP), and functional integrations of the QCSP. The investigation of the individual integration phases is subject of the major studies of the thesis.

A major study concerning the BACAP is a third scientific contribution of the thesis. This combined problem of berth allocation and crane assignment has been formulated first by Park and Kim (2003). A new mathematical formulation is provided in the thesis. The model respects crane productivity losses effected by crane interference and by berthing vessels apart from desired berthing positions. The effects have been modeled using a so-called interference exponent and a berth deviation factor, respectively. Despite its extensions, the new mathematical formulation is more compact than the model of Park and Kim (2003). For solving the BACAP, several heuristic solution methods have been presented. The most important findings of

F. Meisel, *Seaside Operations Planning in Container Terminals*,
Contributions to Management Science,
DOI: 10.1007/978-3-7908-2191-8_9,

the computational tests are the following: First, productivity decreasing effects have a strong influence on the solution quality and, therefore, should be taken into account when berth allocation and crane assignment are solved in practice. Second, the proposed solution methods are effective in pursuing sufficient service quality and low operational cost. Finally, and most relevant for the aspired integration approach, the quality of solutions generated for the BACAP clearly dominates solutions obtained from a sequential planning of berth allocation and crane assignment.

Taking up a current stream of research, the thesis furthermore contributes a new solution approach for the QCSP with crane tasks defined by container groups. For this problem, a set of new constraints is presented in order to avoid crossings of cranes and violations of safety margins in a schedule. A heuristic method is provided, which searches the space of so-called unidirectional schedules. Computational tests reveal that this heuristic clearly outperforms the methods proposed in the field in terms of solution quality and computation time. Near optimal solutions are found quickly even for large-sized instances which have not been tackled previously. It has been shown that better QCSP solutions are obtained if tasks are defined by container groups instead of defining them by complete bays or by bay areas. Safety margins have a strong impact on the makespan of a crane schedule, which necessitates their consideration within QCSP formulations. Furthermore, the QCSP model and the heuristic have been extended towards the Quay Crane Scheduling Problem with Time Windows (QCSPTW). In this problem, the availability of cranes at a vessel is restricted by given time windows. Although solutions generated by the adapted heuristic show a considerable gap, acceptable solution quality is achieved within a restricted runtime of 1 min.

The integration of crane scheduling into the BACAP is a fifth scientific contribution of the thesis. A preprocessing of crane schedules provides a crane productivity estimate to the BACAP in order to support the assignment of sufficient crane capacity to vessels. Computational tests show that this crane productivity information considerably improves the BACAP solutions. A conducted evaluation has revealed that the solutions require only little revision once final crane schedules are generated for the vessels. The alternative of using an interference exponent for crane productivity estimation has shown to be not competitive. Hence, it is concluded that seaside operations planning benefits from the integration of the QCSP into the BACAP by a preprocessing. Furthermore, a feedback loop integration of the QCSPTW into the berth planning has been investigated. This integration phase ensures that a vessel receives enough crane capacity for a complete service. Possible resource conflicts are resolved by a repair process. While the BACAP ensures that the interrelations of berth planning and crane assignment are respected, the feedback loop further ensures that the interrelations with the crane scheduling decisions are respected. The crane schedules obtained within the feedback loop phase complete the outcome of the seaside planning process.

Summarizing the thesis, three important container terminal planning problems have been investigated. The provided mathematical formulations and solution methods allow to obtain solutions of good or even optimal quality in reasonable computation times. All phases of the concept for the integrated solution of the problems have shown to improve seaside operations planning. Applying the integration concept enables a CT management to obtain feasible plans for the seaside operations where berthing positions, berthing times, crane assignments, and crane schedules are consistently determined for the vessels. The obtained plans enable a highly productive utilization of quay space and quay cranes as well as a reliable service of vessels within the projected handling time intervals.

Despite of the achieved progress, seaside operations planning offers further research potential. Enhancements are possible for modeling and solving each single optimization problem. For example, more powerful heuristics can be developed for the BACAP and the QCSPTW. An open research issue regarding the QCSP with container groups is to identify characteristics when optimality comes along with unidirectionality. On a broader scope, seaside operations planning can be viewed more holistically by extending the integration concept. For example, stability issues can be incorporated into QCSP formulations through an integration of stowage planning. Furthermore, the crane productivity loss caused by apart berthing of vessels is modeled through a constant berth deviation factor. The effect can be investigated in detail if the planning of container transports is integrated, too. The presented classification schemes for the seaside planning problems can help to identify further research potential.

Appendix A

A.1 The Berth Allocation and Crane Assignment Model of Park and Kim (2003)

The berth allocation and crane assignment model of Park and Kim (2003) is given below. To ease comparison with the BACAP model presented in sect. 6.1 the denotation used by Park and Kim is replaced by the denotation used in the thesis. The following redefinitions and additional variables are required, see Park and Kim (2003):

T Set of 1-h periods, $T = \{1, \ldots, H\}$, H is the planning horizon
c_i^1 Cost of vessel i per unit distance of berthing apart from the desired position
c_i^2 Cost of vessel i per unit time of arrival before ETA_i (speedup cost)
c_i^3 Cost of vessel i per unit time of arrival after ETA_i (waiting cost)
c_i^4 Cost of vessel i per unit time of delay beyond EFT_i (tardiness cost)
q_{it} Integer decision variable, number of cranes assigned to vessel i in period t
X_{bti} Binary decision variable, set to 1 if the space–time segment (b,t) is covered by the rectangle of vessel i, 0 otherwise
Z_{bti} Binary decision variable, set to 1 if the rectangle of vessel i is located at (b,t) in the space–time diagram, 0 otherwise

The berth allocation and crane assignment model of Park and Kim (2003):

$$\text{minimize } Z = \sum_{i=1}^{n} \sum_{b=1}^{L} \sum_{t=1}^{H} Z_{bti} \left\{ c_i^1 |b - b_i^0| + c_i^2 (ETA_i - t)^+ + c_i^3 (t - ETA_i)^+ + c_i^4 (e_i - EFT_i)^+ \right\} \quad \text{(A.1)}$$

subject to

$$\sum_{i=1}^{n} X_{bti} \le 1 \qquad \forall b \in \{1, \ldots L\}, \ \forall t \in T, \quad \text{(A.2)}$$

F. Meisel, *Seaside Operations Planning in Container Terminals*, Contributions to Management Science,
DOI: 10.1007/978-3-7908-2191-8_A

$$\sum_{i=1}^{n} q_{it} \leq Q \qquad \forall t \in T, \tag{A.3}$$

$$\sum_{t=1}^{H} q_{it} \geq m_i \qquad \forall i \in V, \tag{A.4}$$

$$q_{it} \leq M \cdot r_{it} \qquad \forall t \in T, \forall i \in V, \tag{A.5}$$

$$r_{it} \leq q_{it} \qquad \forall t \in T, \forall i \in V, \tag{A.6}$$

$$(t+1) \cdot r_{it} \leq e_i \qquad \forall t \in T, \forall i \in V, \tag{A.7}$$

$$(t'' - t' + 1) \leq \sum_{t=t'}^{t''} r_{it} + M(2 - r_{it'} - r_{it''}) \qquad \forall t', t'' \in T, t' < t'', \forall i \in V, \tag{A.8}$$

$$r_{it} \leq \sum_{b=1}^{L} \sum_{t'=1}^{t} Z_{bt'i} \qquad \forall t \in T, \forall i \in V, \tag{A.9}$$

$$\sum_{b=1}^{L} \sum_{t=1}^{H} Z_{bti} = 1 \qquad \forall i \in V, \tag{A.10}$$

$$\sum_{b=1}^{b'-1} \sum_{t=1}^{H} X_{bti} + \sum_{b=b'+l_i}^{L} \sum_{t=1}^{H} X_{bti} \leq M\left(1 - \sum_{t=1}^{H} Z_{b'ti}\right) \qquad \forall b' \in \{2, \ldots L - l_i\}, \forall i \in V, \tag{A.11}$$

$$\sum_{b=1+l_i}^{L} \sum_{t=1}^{H} X_{bti} \leq \left(1 - \sum_{t=1}^{H} Z_{1ti}\right) \qquad \forall i \in V, \tag{A.12}$$

$$\sum_{b=1}^{L-l_i} \sum_{t=1}^{H} X_{bti} \leq \left(1 - \sum_{t=1}^{H} Z_{(L-l_i+1)ti}\right) \qquad \forall i \in V, \tag{A.13}$$

$$l_i - \sum_{b=1}^{L} X_{bti} \leq M(1 - r_{it}) \qquad \forall t \in T, \forall i \in V, \tag{A.14}$$

$$X_{bti}, Z_{bti}, r_{it} \in \{0, 1\}, \tag{A.15}$$

$$q_{it} \in \{0, r_i^{\min}, \ldots r_i^{\max}\}. \tag{A.16}$$

For a description of the model see Park and Kim (2003).

Appendix B

B.1 Pseudocodes

Algorithm B.1 Crane assignment procedure used in the BACAP construction heuristic

1: **procedure** QC_ASSIGNMENT(i, s_i, b_i)
2: **for each** $t \in T$ **do** $Q_t :=$ number of unassigned QCs in period t;
3: **for each** $t \in T, q \in R_i$ **do** $r_{itq} := 0$; ▷ Initialize assignment.
4: $e_i := s_i + d_i^{\min}$; ▷ See (6.1) for $d_i^{\min}$.
5: **if** $(e_i > H)$ or $(\exists t \in [s_i, e_i - 1] : Q_t < r_i^{\min})$ ▷ No feasible assignment ...
6: **then** return false; ▷ ... possible → terminate.

▷ Assign crane capacity within minimum handling interval.
7: **for each** $t \in [s_i, e_i - 1]$ **do**
8: $q := min(Q_t, r_i^{\max})$; ▷ Determine number of QCs to assign.
9: $r_{itq} := 1$; ▷ Assign QCs.
10: **end for**

▷ While assigned crane capacity is insufficient: extend the handling interval.
11: **while** not(Constraint (6.3) holds for vessel i) **do**
12: $e_i := e_i + 1$;
13: **if** $(e_i > H)$ or $(Q_{e_i - 1} < r_i^{\min})$ **then** return false; ▷ Assignment impossible.
14: $q := min(Q_{e_i - 1}, r_i^{\max})$;
15: $r_{i, e_i - 1, q} := 1$;
16: **end while**

▷ Realize an almost uniform crane assignment within the handling interval.
17: **for each** $t \in T, q \in R_i$ **do** $r_{itq} := 0$; ▷ Reset assignment.
18: **for** $q := r_i^{\min}$ **to** $r_i^{\max}$ **do** ▷ Increment the assigned QCs ...
19: **for** $t := s_i$ **to** $e_i - 1$ **do** ▷ ... in the periods of the handling interval.
20: **if** $q \leq Q_t$ **then**
21: $r_{itq'} := 0 \ \forall q' \in R_i$; ▷ Reset assignment at period t.
22: $r_{itq} := 1$; ▷ Assign q QCs.
23: **if** Constraint (6.3) holds for vessel i **then** goto (27);
24: **end if**
25: **end for**
26: **end for**

27: Return true; ▷ Feasible assignment found.
28: **end procedure**

F. Meisel, *Seaside Operations Planning in Container Terminals*,
Contributions to Management Science,
DOI: 10.1007/978-3-7908-2191-8_B © Physica-Verlag Berlin Heidelberg 2009

Algorithm B.2 Local refinement by resource leveling

procedure RESOURCE_LEVELING(current solution *BerthPlan*, priority list *P*)
  bestBerthPlan := *BerthPlan*; ▷ The so far best known solution.
  Remove all vessels from *BerthPlan*; ▷ Unschedule the vessels.
  for each $i \in V$ **do** $r_i^{\text{lvl}} := r_i^{\max}$; ▷ Initialize resource levels.

  for $i := 1$ **to** $n-1$ **do**
    for each $r \in R_{p_i}$ **do**
      $r_{p_i}^{\text{lvl}} := r$; ▷ Apply resource restriction to vessel p_i.
      $Insert(p_i)$;
      for $j := i+1$ **to** n **do** $Insert(p_j)$; ▷ Schedule low priority vessels.

      Remove vessel p_i from *BerthPlan*; ▷ Reinsert vessel p_i ...
      $r_{p_i}^{\text{lvl}} := r_{p_i}^{\max}$; ▷ ... without ...
      $Insert(p_i)$; ▷ ... resource restriction.
      $r_{p_i}^{\text{lvl}} := r$; ▷ Restore resource restriction.

      if Cost(*BerthPlan*) < Cost(*bestBerthPlan*) **then**
        bestBerthPlan := *BerthPlan*; ▷ Store new best solution.
        Store resource level $r_{p_i}^{\text{lvl}}$;
      end if
    end for
    Restore partial solution of vessels p_1 to p_i from *bestBerthPlan*;
  end for

  Restore *bestBerthPlan* and corresponding resource levels;
end procedure

Algorithm B.3 Local refinement by vessel shifting

```
1:  procedure VESSEL_SHIFTING(current solution BerthPlan)
2:    bestBerthPlan := BerthPlan;                          ▷ The so far best known solution.

3:    repeat
4:      Identify spatial clusters and store them in spatialClusters;

        ▷ Shift every spatial cluster to lower quay border.
5:      for each Cluster ∈ spatialClusters do
6:        max_seg := max(b_i | i ∈ V);                     ▷ Maximum number of segments to shift.
7:        for max_seg times do

8:          for each i ∈ Cluster do b_i := b_i − 1;         ▷ Shift vessels.
9:          for each i ∈ Cluster do                         ▷ Check vessels.
10:           if b_i < b_i^0 then                           ▷ Reinsert vessel i if shifted away from b_i^0 ...
11:             Remove vessel i from BerthPlan;             ▷ ... (avoids also ...
12:             Insert(i);                                  ▷ ... leaving the quay area).
13:           else
14:             Call QC_ASSIGNMENT(i, s_i, b_i);            ▷ Adapt QC assignment.
15:             if (assignment failed) or (vessel rectangles overlapp) then
16:               Remove vessel i from BerthPlan;           ▷ Repair infeasible solution ...
17:               Insert(i);                                ▷ ... by reinsertion.
18:           end if
19:         end for

20:         for each i ∈ Cluster do
21:           if r_i^lvl ≠ r_i^max then                     ▷ cf. lines 10 to 13 of procedure ...
22:             Relax resource level of i;                  ▷ ... RESOURCE_LEVELING.

23:         if Cost(BerthPlan) < Cost(bestBerthPlan) then
24:           bestBerthPlan := BerthPlan;                   ▷ Store new best solution.

25:         if All vessels in Cluster required reinsertion within last loop then
26:           goto (29);                                    ▷ End shifting of cluster.

27:         Update Cluster;                                 ▷ Add vessels fulfilling the spatial cluster property.
28:       end for

29:       Restore bestBerthPlan;
30:     end for

        [Perform similar operations to shift spatial clusters to upper quay border ...
        ... and to shift temporal clusters towards time 0 and time H, respectively.]

31:   until No improvement found within last loop iteration;
32:   Restore bestBerthPlan;
33: end procedure
```

Algorithm B.4 Squeaky wheel optimization

```
procedure SWO(current solution BerthPlan, priority list P, iterations max_iter)
    bestBerthPlan := BerthPlan;                          ▷ The so far best known solution.
    P_bestBerthPlan := P;                                ▷ The corresponding priority list.
    investigatedPriorityLists := ∅;                      ▷ The already investigated priority lists.
    iter := 0;                                           ▷ Iteration counter.
    iter_best := 0;                                      ▷ Iteration of best solution.
    for each i ∈ V do TSC_i := 0;                        ▷ Total service quality cost (TSC) of vessel i.

    repeat
        iter := iter + 1;
        Construct a solution from P and store it as BerthPlan;      ▷ Construct.

        if P ∉ investigatedPriorityLists then                       ▷ Avoids refinement ...
            Call RESOURCE_LEVELING(BerthPlan, P);                   ▷ ... in case of...
            Call VESSEL_SHIFTING(BerthPlan);                        ▷ ... cycling.
        end if

        if Cost(BerthPlan) < Cost(bestBerthPlan) then
            bestBerthPlan := BerthPlan;                             ▷ Store new best solution.
            iter_best := iter;
            P_bestBerthPlan := P;
        end if

        investigatedPriorityLists := investigatedPriorityLists ∪ P;

    ▷ Analyze solution by determining total service quality cost of each vessel.
        for each i ∈ V do
            TSC_i := TSC_i + c_i^1 · ΔETA_i + c_i^2 · ΔEFT_i + c_i^3 · u_i;

    ▷ Prioritize vessels.
        for i := 1 to n-1 do
            if TSC_{p_i} < TSC_{p_{i+1}} then
            Swap p_i and p_{i+1} in priority list P;

    until iter > iter_best + max_iter;                   ▷ Stop after max_iter iterations ...
                                                         ▷ ... without improvement.

    Restore bestBerthPlan and corresponding priority list P_bestBerthPlan;
end procedure
```

Algorithm B.5 Tabu search

```
procedure TS(current solution BerthPlan, priority list P, iterations max_iter)
    bestBerthPlan := BerthPlan;                         ▷ The so far best known solution.
    P_bestBerthPlan := P;                               ▷ The corresponding priority list.
    tabuList := ∅;                                      ▷ The tabu list.
    iter := 0;                                          ▷ Iteration counter.
    iter_best := 0;                                     ▷ Iteration of best solution.

    repeat
        iter := iter + 1;
        bestNeighbor := ∅;
        cost_bestNeighbor := ∞;
        P_bestNeighbor := ();

        for i := 1 to n − 1 do                          ▷ Evaluate ...
            for j := i + 1 to n do                      ▷ ... neighborhood.
                Exchange p_i and p_j in priority list P;
                Construct a solution from P and store the solution as Neighbor;
                if Neighbor ∉ tabuList then
                    if Cost(Neighbor) < cost_bestNeighbor then
                        bestNeighbor := Neighbor;       ▷ Store neighbor.
                        cost_bestNeighbor := Cost(bestNeighbor);
                        P_bestNeighbor := P;
                    end if
                end if
                Exchange p_i and p_j in priority list P; ▷ Restore priority list.
            end for
        end for

        BerthPlan := bestNeighbor;                      ▷ Restore best performing neighbor ...
        P := P_bestNeighbor;                            ▷ ... and corresponding priority list.
        tabuList := tabuList ∪ BerthPlan;
        Call RESOURCE_LEVELING(BerthPlan, P);
        Call VESSEL_SHIFTING(BerthPlan);

        if Cost(BerthPlan) < Cost(bestBerthPlan) then
            bestBerthPlan := BerthPlan;
            iter_best := iter;
            P_bestBerthPlan := P;
        end if

    until iter > iter_best + max_iter;                  ▷ Stop after max_iter iterations ...
                                                        ▷ ... without improvement.

    Restore bestBerthPlan and corresponding priority list P_bestBerthPlan;
end procedure
```

Algorithm B.6 QCSP schedule generation scheme (determines task completion times in a unidirectional schedule corresponding to a given task-to-QC assignment)

```
procedure QCSP_SCHEDULE_GENERATOR( task-to-QC assignment m,
Disj. Graph G = (Ω̅, A, D, W))
    c_0 := 0;

    for k := q downto 1 do                          ▷ QCs in inverse order (priority to upper QC).
        for i := 1 to n do                          ▷ Schedule tasks which are ...
            if m_i = k then                         ▷ ... assigned to QC k.

                if (0, i) ∈ A_k then                ▷ First in task sequence of QC k.
                    c_i := r^k + t^k_{0i} + p_i;
                else
                    for j := 1 to i − 1 do          ▷ Find preceding task ...
                        if (j, i) ∈ A_k then        ▷ ... in sequence of QC k.
                            c_i := c_j + t_{ji} + p_i;

                for each j ∈ Ω do                   ▷ Other tasks which are assigned ...
                    if m_j > k then                 ▷ ... to QCs with higher priority.
                        if (j, i) ∈ D ∪ A_Φ then    ▷ An arc exists in the graph.
                            c_i := max(c_i, c_j + Δ^{m_j,k}_{j,i} + p_i);   ▷ Delay task if required.
            end if
        end for
    end for

    c_T := max(c_i | i ∈ Ω);                        ▷ Determine makespan.
end procedure
```

Algorithm B.7 QCSPTW schedule generation scheme

```
1: procedure QCSPTW_SCHEDULE_GENERATOR( task-to-QC assignment m,
   Disj. Graph G = (Ω̄, A, D, W) )
2:     c_0 := 0;

3:     for k := q downto 1 do                  ▷ QCs in inverse order (priority to upper QC).
4:         for i := 1 to n do                  ▷ Schedule tasks which are ...
5:             if m_i = k then                 ▷ ... assigned to QC k.

6:                 ... [lines 6 to 15 of QCSP_SCHEDULE_GENERATOR] ...

7:                 for u := 1 to τ_k do        ▷ Investigate time windows of QC k.

8:                     if c_i − p_i < r^ku + t^ku_0i then   ▷ If task starts too early ...
9:                         c_i := r^ku + t^ku_0i + p_i;     ▷ ... postpone task.

10:                    if c_i + t^ku_iT ≤ d^ku then         ▷ If processing ends within time window ...
11:                        goto (4);                        ▷ ... continue with next task.

12:                    if u = τ_k then          ▷ If no appropriate time windows has been found ...
13:                        return false;        ▷ ... the schedule generation failed.

14:                end for

15:            end if
16:        end for
17:    end for

18:    c_T := max(c_i | i ∈ Ω);                ▷ Determine makespan.
19:    Return true;                            ▷ Feasible schedule found.
20: end procedure
```

Appendix C

C.1 A Lower Bound for the QCSP

A mathematical model is provided for determining a lower bound on the minimum makespan of a QCSP instance. In order to enable ILOG CPLEX to solve large-sized instances of the model, the primary decision made is the assignment of tasks to QCs. Scheduling issues such as determination of completion times for the tasks are not considered. To strengthen the bound, the least required crane movement time is estimated as follows. Let B be the number of bays that contain at least one task but where no QC is initially positioned. Obviously, moving QCs to these bays requires at least $B \cdot \hat{t}$ time units. In the model, this movement time can be shared arbitrarily among QCs in order to keep the formulation solvable. Next to the denotation introduced in Sect. 7.1, the following decision variables are introduced:

- y_i^k Binary, set to 1 if task $i \in \Omega$ is assigned to QC $k \in Q$, 0 otherwise
- z^k Binary, set to 1 if QC k processes at least one task, 0 otherwise
- t^k Integer, movement time assigned to QC k

A lower bound on the optimal makespan of a QCSP instance follows from solving:

$$\text{minimize } c_T \tag{C.1}$$

subject to

$$c_T \geq z^k r^k + \sum_{i \in \Omega} y_i^k p_i + t^k \quad \forall k \in Q, \tag{C.2}$$

$$\sum_{k \in Q} y_i^k = 1 \quad \forall i \in \Omega, \tag{C.3}$$

$$\sum_{k \in Q} t^k = B\hat{t} \tag{C.4}$$

F. Meisel, *Seaside Operations Planning in Container Terminals*,
Contributions to Management Science,
DOI: 10.1007/978-3-7908-2191-8_C © Physica-Verlag Berlin Heidelberg 2009

$$\sum_{i \in \Omega} y_i^k \leq Mz^k \qquad \forall k \in Q, \tag{C.5}$$

$$y_i^k, z^k \in \{0,1\} \qquad \forall i \in \Omega, \quad \forall k \in Q, \tag{C.6}$$

$$t^k \geq 0 \qquad \forall k \in Q. \tag{C.7}$$

The pursued objective in (C.1) is to minimize the completion time of task T. It is determined by the maximum QC finishing time in Constraints (C.2), where the finishing time of a crane is determined by its ready time, the processing time of assigned tasks, and the assigned movement time. Constraints (C.3) ensure that every task is assigned to one QC. Constraint (C.4) distributes the least required movement time to the QCs. Constraints (C.5) set variables z^k to 1 if crane k processes at least one task. Constraints (C.6) and (C.7) define the domains of the decision variables.

Appendix D

D.1 A Lower Bound for the QCSPTW

To obtain a lower bound on the minimum makespan of a QCSPTW instance, the following formulation is solved by ILOG CPLEX. Next to the denotation introduced in Sects. 7.1 and 7.3, the binary decision variables z^{ku} are introduced, set to 1 if QC k processes at least one task in time window u, 0 otherwise.

A lower bound on the optimal makespan of a QCSPTW instance follows from solving:

$$\text{minimize } c_T \tag{D.1}$$

subject to

$$c_T \geq z^{ku} r^{ku} + \sum_{i\in\Omega} y_i^{ku} p_i \qquad \forall k \in Q,\ \forall u \in TW_k, \tag{D.2}$$

$$\sum_{k\in Q}\sum_{u\in TW_u} y_i^{ku} = 1 \qquad \forall i \in \Omega, \tag{D.3}$$

$$\sum_{i\in\Omega} y_i^{ku} p_i \leq d^{ku} - r^{ku} - \left| l_T^{ku} - l_0^{ku} \right| \hat{t} \qquad \forall k \in Q,\ \forall u \in TW_k, \tag{D.4}$$

$$\sum_{i\in\Omega} y_i^{ku} \leq M z^{ku} \qquad \forall k \in Q,\ \forall u \in TW_k, \tag{D.5}$$

$$y_i^{ku}, z^{ku} \in \{0,1\} \qquad \forall i \in \Omega,\ \forall k \in Q,\ \forall u \in TW_k. \tag{D.6}$$

The pursued objective in (D.1) is to minimize the completion time of task T. It is determined by the maximum QC finishing time in Constraints (D.2). Constraints (D.3) ensure that every task is assigned to one time window of a crane. The right side of Constraint (D.4) yields an upper bound on the available time within a time window that can be effectively used by a crane. The constraint ensures that the total processing time of tasks assigned to a time window does not overshoot this capacity. Constraints (D.5) set variables z^{ku} to 1 if crane k processes at least one task in time window u. Constraints (D.6) define the domains of the decision variables.

F. Meisel, *Seaside Operations Planning in Container Terminals*,
Contributions to Management Science,
DOI: 10.1007/978-3-7908-2191-8_D, © Physica-Verlag Berlin Heidelberg 2009

Bibliography

Ak A, Erera AL (2006) Simultaneous berth and quay crane scheduling for container ports, working paper, H. Milton Stewart School of Industrial and Systems Engineering, Georgia Institute of Technology, Atlanta

Atamtürk A, Savelsbergh MWP (2005) Integer-programming software systems. Annals of Operations Research 140(1):67–124

Baird AJ (2006) Optimising the container transhipment hub location in northern europe. Journal of Transport Geography 14(3):195–214

Baker BS, Coffman EGJ, Rivest RL (1980) Orthogonal packings in two dimensions. SIAM Journal on Computing 9(4):846–855

Bierwirth C, Meisel F (2009) A fast heuristic for quay crane scheduling with interference constraints, Journal of Scheduling, doi: 10.1007/s10951-009-0105-0

Bish EK (2003) A multiple-crane-constrained scheduling problem in a container terminal. European Journal of Operational Research 144(1):83–107

Böse J, Reiners T, Steenken D, Voß S (2000) Vehicle dispatching at seaport container terminals using evolutionary algorithms. In: Proceedings of the 33th Hawaii International Conference on System Sciences (HICSS 33), vol 2, pp 1–10

Briano C, Briano E, Bruzzone AG (2005) Models for support maritime logistics: a case study for improving terminal planning. In: Merkuryev Y, Zobel R, Kerckhoffs E (eds) Proceedings of the 19th European Conference on Modelling and Simulation (ECMS), pp 199–203

Brinkmann B (2005) Seaports – planning and design (in German). Springer, Berlin

Brown GG, Lawphongpanich S, Thurman KP (1994) Optimizing ship berthing. Naval Research Logistics 41(1):1–15

Brown GG, Cormican KJ, Lawphongpanich S, Widdis D (1997) Optimizing submarine berthing with a persistence incentive. Naval Research Logistics 44(4):301–318

Canonaco P, Legato P, Mazza RM, Musmanno R (2008) A queuing network model for the management of berth crane operations. Computers and Operations Research 35(8):2432–2446

Chen CY, Hsieh TW (1999) A time–space network model for the berth allocation problem. 19th IFIP TC7 Conference on System Modeling and Optimization, Cambridge

Chen L, Bostel N, Dejax P, Cai J, Xi L (2007) A tabu search algorithm for the integrated scheduling problem of container handling systems in a maritime terminal. European Journal of Operational Research 181(1):40–58

Cheong CY, Lin CJ, Tan KC, Liu DK (2007) A multi-objective evolutionary algorithm for berth allocation in a container port. IEEE Computer Society, Washington DC, IEEE Congress on Evolutionary Computation 2007 (CEC 2007), pp 927–934

Cheung RK, Li CL, Lin W (2002) Interblock crane deployment in container terminals. Transportation Science 36(1):79–93

Chu CY, Huang WC (2002) Aggregates cranes handling capacity of container terminals: the port of kaohsiung. Maritime Policy and Management 29(4):341–350

Clements DP, Crawford JM, Joslin DE, Nemhauser GL, Puttlitz ME, Savelsbergh MWP (1997) Heuristic optimization: a hybrid AI/OR approach. In: Proceedings of the Workshop on Industrial Constraint-Directed Scheduling, in conjunction with the Third International Conference on Principles and Practice of Constraint Programming (CP97)

Cordeau JF, Laporte G, Legato P, Moccia L (2005) Models and tabu search heuristics for the berth-allocation problem. Transportation Science 39(4):526–538

Crainic TG, Kim KH (2007) Intermodal transportation. Handbooks in Operations Research and Management Science, vol 14. Elsevier, Amsterdam, pp 467–537

Daganzo CF (1989) The crane scheduling problem. Transportation Research Part B 23(3):159–175

Dai J, Lin W, Moorthy R, Teo CP (2008) Berth allocation planning optimization in container terminals. In: Tang CS, Teo CP, Wei KK (eds) Supply chain analysis: a handbook on the interaction of information, system and optimization. Springer, New York, pp 69–105

Dekker R, Voogd P, van Asperen E (2006) Advanced methods for container stacking. OR Spectrum 28(4):563–586

Dolk DR, Kottemann JE (1993) Model integration and a theory of models. Decision Support Systems 9(1):51–63

Dragovic B, Park NK, Radmilovic Z, Maras V (2005) Simulation modelling of ship-berth link with priority service. Maritime Economics and Logistics 7(4):316–335

Dragovic B, Park NK, Radmilovic Z (2006) Ship-berth link performance evaluation: simulation and analytical approaches. Maritime Policy and Management 33(3):281–299

Edmond ED, Maggs RP (1978) How useful are queue models in port investment decisions for container berths. Journal of the Operational Research Society 29(8):741–750

Froyland G, Koch T, Megow N, Duane E, Wren H (2008) Optimizing the landside operation of a container terminal. OR Spectrum 30(1):53–75

Fu Z, Li Y, Lim A, Rodrigues B (2007) Port space allocation with a time dimension. Journal of the Operational Research Society 58(6):797–807

Gambardella LM, Mastrolilli M, Rizzoli AE, Zaffalon M (2001) An optimization methodology for intermodal terminal management. Journal of Intelligent Manufacturing 12(5):521–534

Geoffrion A (1999) Structured modeling: survey and future research directions. Interactive Transactions of ORMS 1(2)

Giallombardo G, Moccia L, Salani M, Vacca I (2008) The tactical berth allocation problem with quay crane assignment and transshipment-related quadratic yard costs. In: Proceedings of the European Transport Conference (ETC), pp 1–27

Glover F (1986) Future paths for integer programming and links to artificial intelligence. Computers and Operations Research 13(5):533–549

Goedhart GJ (2002) Criteria for (un)-loading container ships. Technical report, Technical University Delft, Netherlands

Goh KS, Lim A (2000) Combining various algorithms to solve the ship berthing problem. In: Proceedings of the 12th IEEE International Conference on Tools with Artificial Intelligence (ICTAI'00), IEEE Computer Society, Los Alamitos, CA, pp 370–373

Golias M, Boile M, Theofanis S (2006) The berth allocation problem: a formulation reflecting time window service deadlines. In: Proceedings of the 48th Transportation Research Forum Annual Meeting, Transportation Research Forum, Boston

Golias M, Boile M, Theofanis S (2007) The stochastic berth allocation problem. In: Proceedings of the International Conference on Transport Science and Technology (TRANSTEC 2007), Czech Technical University, Prague, pp 52–66

Goodchild AV (2006) Port planning for double cycling crane operations. In: Proceedings of the 85th Annual Meeting of Transportation Research Board (CD-ROM), Washington DC, Annual Meeting of Transportation Research Board

Goodchild AV, Daganzo CF (2004) Reducing ship turn-around time using double-cycling, research report UCB-ITS-RR-2004-4, University of California, Berkeley

Goodchild AV, Daganzo CF (2005a) Crane double cycling in container ports: affect on ship dwell time, research report RR20055, University of California, Berkeley

Goodchild AV, Daganzo CF (2005b) Performance comparison of crane double-cycling strategies, working paper UCB-ITS-WP-2005-2, University of California, Berkeley

Goodchild AV, Daganzo CF (2006) Double-cycling strategies for container ships and their effect on ship loading and unloading operations. Transportation Science 40(4):473–483

Goodchild AV, Daganzo CF (2007) Crane double cycling in container ports: planning methods and evaluation. Transportation Research Part B 41(8):875–891

Gottwald Port Technology (2008) Photo gallery. Düsseldorf, http://www.gottwald.com/gottwald/site/gottwald/de/news/gallery/. Accessed 9 Apr 2008

Grunow M, Günther HO, Lehmann M (2006) Strategies for dispatching AGVs at automated seaport container terminals. OR Spectrum 28(4):587–610

Guan Y, Cheung RK (2004) The berth allocation problem: models and solution methods. OR Spectrum 26(1):75–92

Guan Y, Xiao WQ, Cheung RK, Li CL (2002) A multiprocessor task scheduling model for berth allocation: heuristic and worst-case analysis. Operations Research Letters 30(5):343–350

Günther HO, Kim KH (eds) (2005) Container terminals and automated transport systems. Springer, Berlin

Han M, Li P, Sun J (2006) The algorithm for berth scheduling problem by the hybrid optimization strategy GASA. In: Proceedings of the 9th International Conference on Control, Automation, Robotics and Vision (ICARCV '06), IEEE Computer Society, Washington DC, pp 1–4

Hansen P, Oğuz C (2003) A note on formulations of static and dynamic berth allocation problems. Les Cahiers du GERAD 30:1–17

Hansen P, Oğuz C, Mladenovic N (2008) Variable neighborhood search for minimum cost berth allocation. European Journal of Operational Research 191(3):636–649

Hapag-Lloyd (2008) Containers. http://www.hapag-lloyd.com/en/fleet/container.html. Accessed 9 Apr 2008

Hartmann S (2004) A general framework for scheduling equipment and manpower at container terminals. OR Spectrum 26(1):51–74

Hax AC, Meal D (1975) Hierarchical integration of production planning and scheduling. In: Geisler MA (ed) Logistics, TIMS studies in the management sciences. North Holland, Amsterdam, pp 53–69

Hendriks MPM, Laumanns M, Lefeber E, Udding JT (2008) Robust periodic berth planning of container vessels. In: Kopfer H, Günther HO, Kim KH (eds) Proceedings of the 3rd German-Korean Workshop on Container Terminal Management: IT-based Planning and Control of Seaport Container Terminals and Transportation Systems, pp 1–13

Henesey L, Davidsson P, Persson JA (2004) Using simulation in evaluating berth allocation at a container terminal. Presented at 3rd International Conference on Computer Applications and Information Technology in the Maritime Industries (COMPIT'04), Siguënza, May 9–12

HHLA (2008) Chronicle. Hamburger Hafen und Logistik AG, Hamburg, http://www.hhla.de/chronicle.53.0.html. Accessed 9 Apr 2008

Hoffarth L, Voß S (1994) Berth allocation in a container terminal – development of a decision support system (in German). In: Dyckhoff H, Derigs U, Salomon M, Tijms HC (eds) Operations Research Proceedings 1993, Springer, Berlin, pp 89–95

IATA (2007) ULD technical manual, 22nd edn. International Air Transport Association, Montreal

Ilmer M (2005) Beating congestion by building capacity: an overview of new container terminal developments in Northern Europe. Port Technology International PT28-06/2:1–5

Imai A, Nagaiwa K, Tat CW (1997) Efficient planning of berth allocation for container terminals in asia. Journal of Advanced Transportation 31(1):75–94

Imai A, Nishimura E, Papadimitriou S (2001) The dynamic berth allocation problem for a container port. Transportation Research Part B 35(4):401–417

Imai A, Nishimura E, Papadimitriou S (2003) Berth allocation with service priority. Transportation Research Part B 37(5):437–457

Imai A, Sun X, Nishimura E, Papadimitriou S (2005) Berth allocation in a container port: using a continuous location space approach. Transportation Research Part B 39(3):199–221

Imai A, Sasaki K, Nishimura E, Papadimitriou S (2006) Multi-objective simultaneous stowage and load planning for a container ship with container rehandle in yard stacks. European Journal of Operational Research 171(2):373–389

Imai A, Nishimura E, Hattori M, Papadimitriou S (2007a) Berth allocation at indented berths for mega-containerships. European Journal of Operational Research 179(2):579–593

Imai A, Zhang JT, Nishimura E, Papadimitriou S (2007b) The berth allocation problem with service time and delay time objectives. Maritime Economics and Logistics 9(4):269–290

Imai A, Chen HC, Nishimura E, Papadimitriou S (2008a) The simultaneous berth and quay crane allocation problem. Transportation Research Part E 44(5):900–920

Imai A, Nishimura E, Papadimitriou S (2008b) Berthing ships at a multi-user container terminal with a limited quay capacity. Transportation Research Part E 44(1):136–151

ISL (2003) Algorithms for the capacity determination of a container terminal at the example of CT IV in Bremerhaven (in German). Technical report, Institute of Shipping Economics and Logistics, Bremerhaven, unpublished

ISL (2008) Maritime Market Monitor. Institute of Shipping Economics and Logistics, Bremerhaven, http://www.isl.org/products_services/market_monitor/shipping.shtml. Accessed 9 Apr 2008

Johnson S (1954) Optimal two- and three-stage production schedules with setup times. Naval Research Logistics Quarterly 1(1):61–68

Jordan MA (2002) Quay crane productivity. Presented at TOC 2002 Americas, Miami, November 19–21

Joslin DE, Clements DP (1999) 'Squeaky wheel' optimization. Journal of Artifical Intelligence Research 10:353–373

Jung SH, Kim KH (2006) Load scheduling for multiple quay cranes in port container terminals. Journal of Intelligent Manufacturing 17(4):479–492

Jung DH, Park YM, Lee BK, Kim KH, Ryu KR (2006) A quay crane scheduling method considering interference of yard cranes in container terminals. In: Gelbukh AF, García CAR (eds) Proceedings of the 5th Mexican International Conference on Artificial Intelligence (MICAI 2006), Springer, Berlin, LNCS, vol 4293, pp 461–471

Kalmar Industries (2008) Products / Pressrelease-pictures. http://www.kalmarind.com/show.php?id=605. Accessed 9 Apr 2008

Kim KH (2005) Models and methods for operations in port container terminals. In: Langevin A, Riopel D (eds) Logistics systems: design and optimization, Gerad 25th Anniversary Series. Springer, Berlin, pp 213–243

Kim KH, Moon KC (2003) Berth scheduling by simulated annealing. Transportation Research Part B 37(6):541–560

Kim KH, Park YM (2004) A crane scheduling method for port container terminals. European Journal of Operational Research 156(3):752–768

Kim CW, Tanchoco JMA, Koo PH (1997) Deadlock prevention in manufacturing systems with agv systems: banker's algorithm approach. Journal of Manufacturing Science and Engineering 119(4B):849–854

Kim KH, Lee KM, Hwang H (2003) Sequencing delivery and receiving operations for yard cranes in port container terminals. International Journal of Production Economics 84(3):283–292

Kim KH, Kang JS, Ryu KR (2004a) A beam search algorithm for the load sequencing of outbound containers in port container terminals. OR Spectrum 26(1):93–116

Kim KH, Kim KW, Hwang H, Ko CS (2004b) Operator-scheduling using a constraint satisfaction technique in port container terminals. Computers and Industrial Engineering 46(2):373–381

Kim KH, Jeon SM, Ryu KR (2006) Deadlock prevention for automated guided vehicles in automated container terminals. OR Spectrum 28(4):659–679

Kolisch R (1995) Project scheduling under resource constraints. Physica, Heidelberg

Lai KK, Shih K (1992) A study of container berth allocation. Journal of Advanced Transportation 26(1):45–60

Laine JT, Vepsäläinen APJ (1994) Economies of speed in sea transportation. International Journal of Physical Distribution and Logistics Management 24(8):33–41

Lee Y, Chen CY (2008) An optimization heuristic for the berth scheduling problem. European Journal of Operational Research 196(2):500–508

Lee Y, Hsu NY (2007) An optimization model for the container pre-marshalling problem. Computers and Operations Research 34(11):3295–3313

Lee YH, Kang KR J Ryu, Kim KH (2005) Optimization of container load sequencing by a hybrid of ant colony optimization and tabu search. Lecture Notes in Computer Science, vol 3611. Springer, Berlin, pp 1259–1268

Lee DH, Song L, Wang H (2006) Bilevel programming model and solutions of berth allocation and quay crane scheduling. In: Proceedings of 85th Annual Meeting of Transportation Research Board (CD-ROM), Washington DC, Annual Meeting of Transportation Research Board

Lee DH, Wang HQ, Miao L (2007) An approximation algorithm for quay crane scheduling with non-interference constraints in port container terminals. Presented at Tristan VI, Phuket, June 10–15

Lee DH, Wang HQ, Miao L (2008a) Quay crane scheduling with non-interference constraints in port container terminals. Transportation Research Part E 44(1):124–135

Lee DH, Wang HQ, Miao L (2008b) Quay crane scheduling with handling priority in port container terminals. Engineering Optimization 40(2):179–189

Legato P, Mazza RM (2001) Berth planning and resources optimisation at a container terminal via discrete event simulation. European Journal of Operational Research 133(3):537–547

Legato L, Monaco MF (2004) Human resources management at a marine container terminal. European Journal of Operational Research 156(3):769–781

Legato P, Gulli D, Trunfio R (2008) The quay crane deployment problem at a maritime container terminal. In: Proceedings of the 22th European Conference on Modelling and Simulation (ECMS 2008), pp 53–59

Lehmann M (2006) Operations planning for automated transport systems in seaport container terminals (in German). PhD thesis, Technical University Berlin

Lehmann M, Grunow M, Günther HO (2006) Deadlock handling for real-time control of AGVs at automated container terminals. OR Spectrum 28(4):631–657

Lemper B (1996) The functioning of the market for seaport container transshipment in the North-Range (in German). Institute of Shipping Economics and Logistics, Bremen

Levinson M (2006) The box. Princeton University Press, Princeton

Li CL, Cai X, Lee CY (1998) Scheduling with multiple-job-on-one-processor pattern. IIE Transactions 30(5):433–445

Liang C, Huang Y, Yang Y (2009) A quay crane dynamic scheduling problem by hybrid evolutionary algorithm for berth allocation planning. Computers and Industrial Engineering 56(3):1021–1028. doi: 10.1016/j.cie.2008.09.024

Lim A (1998) The berth planning problem. Operations Research Letters 22(2):105–110

Lim A (1999) An effective ship berthing algorithm. In: Thomas D (ed) Proceedings of the 16th International Joint Conference on Artificial Intelligence (IJCAI-99-Vol-1), Morgan Kaufmann, San Francisco, pp 594–599

Lim A, Rodrigues B, Xiao F, Zhu Y (2002) Crane scheduling using tabu search. In: Proceedings of the 14th IEEE International Conference on Tools with Artificial Intelligence (ICTAI'02), IEEE Computer Society, Washington DC, pp 146–153

Lim A, Rodrigues B, Song L (2004a) Manpower allocation with time windows. Journal of the Operational Research Society 55(11):1178–1186

Lim A, Rodrigues B, Xiao F, Zhu Y (2004b) Crane scheduling with spatial constraints. Naval Research Logistics 51(3):386–406

Lim A, Rodrigues B, Xu Z (2004c) Approximation schemes for the crane scheduling problem. In: Hagerup T, Katajainen J (eds) 9th Scandinavian Workshop on Algorithm Theory (SWAT 2004), Springer, Berlin, LNCS, vol 3111, pp 323–335

Lim A, Rodrigues B, Xu Z (2004d) Solving the crane scheduling problem using intelligent search schemes (extended abstract). In: Wallace M (ed) 10th International Conference on

Principles and Practice of Constraint Programming (CP 2004), LNCS, vol 3258. Springer, Berlin, pp 747–751

Lim A, Rodrigues B, Xu Z (2007) A m-parallel crane scheduling problem with a non-crossing constraint. Naval Research Logistics 54(2):115–127

Linn R, Liu JY, Wan YW, Zhang C, Murty KG (2003) Rubber tired gantry crane deployment for container yard operation. Computers and Industrial Engineering 45(3):429–442

Liu J, Wan YW, Wang L (2006) Quay crane scheduling at container terminals to minimize the maximum relative tardiness of vessel departures. Naval Research Logistics 53(1):60–74

Lokuge P, Alahakoon D (2004) Hybrid BDI agents with improved learning capabilities for adaptive planning in a container terminal application. In: Proceedings of the International Conference on Intelligent Agent Technology (IAT'04), IEEE Computer Society, Los Alamitos, pp 120–126

Lokuge P, Alahakoon D (2005) Reinforcement learning in neuro BDI agents for achieving agent's intentions in vessel berthing applications. In: Proceedings of the 19th International Conference on Advanced Information Networking and Applications (AINA'05), IEEE Computer Society, Los Alamitos, pp 681–686

Lokuge P, Alahakoon D (2007) Improving the adaptability in automated vessel scheduling in container ports using intelligent software agents. European Journal of Operational Research 177(3):1985–2015

Lokuge P, Alahakoon D, Dissanayake P (2004) Collaborative neuro-BDI agents in container terminals. In: Proceedings of the 18th International Conference on Advanced Information Networking and Applications (AINA'04), IEEE Computer Society, Los Alamitos, pp 155–158

Mauri GR, Oliveira ACM, Lorena LAN (2008) A hybrid column generation approach for the berth allocation problem. In: van Hemert J, Cotta C (eds) 8th European Conference on Evolutionary Computation in Combinatorial Optimisation (EvoCOP 2008), LNCS, vol 4972. Springer, Berlin, pp 110–122

Meersmans PJM, Dekker R (2001) Operations research supports container handling. Econometric Institute Report 234, Erasmus University Rotterdam

Meier L, Schumann R (2007) Coordination of interdependent planning systems, a case study. In: Koschke R, Otthein H, Rödiger KH, Ronthaler M (eds) Lecture Notes in Informatics (LNI) P-109. Köllen Druck, Bonn, pp 389–396

Meisel F, Bierwirth C (2006) Integration of berth allocation and crane assignment to improve the resource utilization at a seaport container terminal. In: Haasis HD, Kopfer H, Schönberger J (eds) Operations Research Proceedings 2005. Springer, Berlin, pp 105–110

Meisel F, Bierwirth C (2009) Heuristics for the integration of crane productivity in the berth allocation problem. Transportation Research Part E 45(1):196–209

Meisel F, Wichmann M (2008) Container sequencing for quay cranes with internal reshuffles, working paper, Martin-Luther-University, Halle-Wittenberg

Moccia L, Cordeau JF, Gaudioso M, Laporte G (2006) A branch-and-cut algorithm for the quay crane scheduling problem in a container terminal. Naval Research Logistics 53(1):45–59

Monaco MF, Sammarra M (2007) The berth allocation problem: a strong formulation solved by a lagrangean approach. Transportation Science 41(2):265–280

Moon K (2000) A mathematical model and a heuristic algorithm for berth planning. PhD thesis, Pusan National University, Pusan

Moorthy R, Teo CP (2006) Berth management in container terminal: the template design problem. OR Spectrum 28(4):495–518

Muhanna W, Pick R (1988) Composite models in SYMMS. In: Proceedings 21st Annual Hawaii International Conference on System Sciences (HICSS 21), IEEE Computer Society, Washington DC, vol 3, pp 418–427

Murty KG, Liu J, Wan YW, Linn R (2005a) A decision support system for operations in a container terminal. Decision Support Systems 39(3):309–332

Murty KG, Wan YW, Liu J, Tseng MM, Leung E, Lai KK, Chiu H (2005b) Hongkong international terminals gains elastic capacity using a data-intensive decision-support system. Interfaces 35(1):61–75

Nam KC, Ha WI (2001) Evaluation of handling systems for container terminals. Journal of Waterway, Port, Coastal and Ocean Engineering 127(3):171–175

Ng WC (2005) Crane scheduling in container yards with inter-crane interference. European Journal of Operational Research 164(1):64–78

Ng WC, Mak KL (2006) Quay crane scheduling in container terminals. Engineering Optimization 38(6):723–737

Nishimura E, Imai A, Papadimitriou S (2001) Berth allocation planning in the public berth system by genetic algorithms. European Journal of Operational Research 131(2):282–292

Notteboom T (2006) The time factor in liner shipping service. Maritime Economics and Logistics 8(1):19–39

Oğuz C, Błażewicz J, Cheng TCE, Machowiak M (2004) Berth allocation as a moldable task scheduling problem. In: Proceedings of the 9th International Workshop on Project Management and Scheduling (PMS 2004), Nancy, pp 201–205

Park KT, Kim KH (2002) Berth scheduling for container terminals by using a sub-gradient optimization technique. Journal of the Operational Research Society 53(9):1054–1062

Park YM, Kim KH (2003) A scheduling method for berth and quay cranes. OR Spectrum 25(1):1–23

Peterkofsky RI, Daganzo CF (1990) A branch and bound solution method for the crane scheduling problem. Transportation Research Part B 24(3):159–172

Pinedo M (2002) Scheduling – theory, algorithms, and systems, 2nd edn. Prentice Hall, Englewood Cliffs, NJ

Port of Hamburg Marketing (2008) Port of Hamburg: statistics. Hamburg, http://www.hafen-hamburg.de/content/blogsection/2/33/lang,en/. Accessed 9 Apr 2008

Port of Hamburg Marketing / D Hasenpusch (2008) Port of Hamburg: photos for press. Hamburg, http://www.hafen-hamburg.de/component/option,com_ponygallery/Itemid,30/func,viewcategory/catid,15/lang,en/. Accessed 9 Apr 2008

Pumpe D (2000) A reference model for planning and control of processes in seaport container terminals (in German). Mensch und Buch, Berlin

Rashidi H (2006) Dynamic scheduling of automated guided vehicles. PhD thesis, University of Essex, Colchester

Ronen D (1983) Cargo ships routing and scheduling: survey of models and problems. European Journal of Operational Research 12(2):119–126

Sammarra M, Cordeau JF, Laporte G, Monaco MF (2007) A tabu search heuristic for the quay crane scheduling problem. Journal of Scheduling 10(4–5):327–336

Schneeweiss C (2003) Distributed decision making, 2nd edn. Springer, Berlin

Schonfeld P, Frank S (1984) Optimizing the use of a containership berth. Transportation Research Record 984:56–62

Schonfeld P, Sharafeldien O (1985) Optimal berth and crane combinations in containerports. Journal of Waterway, Port, Coastal and Ocean Engineering 111(6):1060–1072

Schrecker A (2000) Planning and control of automated transport systems (in German). Gabler, Wiesbaden

Silberholz MB, Golden BL, Baker EK (1991) Using simulation to study the impact of work rules on productivity at marine container terminals. Computers and Operations Research 18(5):433–452

Smith TB, Pyle JM (2004) An effective algorithm for project scheduling with arbitrary temporal constraints. In: Proceedings of the 19th National Conference on Artificial Intelligence (AAAI-04), AAAI, Menlo Park, pp 544–549

Stahlbock R, Voß S (2008) Operations research at container terminals: a literature update. OR Spectrum 30(1):1–52

Steenken D, Winter T, Zimmermann UT (2001) Stowage and transport optimization in ship planning. In: Grötschel M, Krumke S, Rambau J (eds) Online optimization of large scale systems. Springer, Berlin, pp 731–745

Steenken D, Voß S, Stahlbock R (2004) Container terminal operation and operations research – a classification and literature review. OR Spectrum 26(1):3–49

Stopford M (1997) Maritime economics, 2nd edn. Routledge, New York

Tavakkoli-Moghaddam R, Makui A, Salahi S, Bazzazi M, Taheri F (2009) An efficient algorithm for solving a new mathematical model for a quay crane scheduling problem in container ports. Computers and Industrial Engineering 56(1):241–248.

Theofanis S, Boile M, Golias M (2007a) An optimization based genetic algorithm heuristic for the berth allocation problem. IEEE Computer Society, Washington DC, IEEE Congress on Evolutionary Computation 2007 (CEC 2007), pp 4439–4445

Theofanis S, Golias M, Boile M (2007b) Berth and quay crane scheduling: a formulation reflecting service deadlines and productivity agreements. In: Proceedings of the International Conference on Transport Science and Technology (TRANSTEC 2007), Czech Technical University, Prague, pp 124–140

Tong CJ, Lau HC, Lim A (1999) Ant colony optimization for the ship berthing problems. In: Thiagarajan PS, Yap R (eds) 5th Asian Computing Science Conference (ASIAN'99), Springer, Berlin, LNCS, vol 1742, pp 359–370

Tozer DR, Penfold A (2001) Ultra-large container ships (ULCS) – designing to the limit of current and projected terminal infrastructure capabilities. In: Proceedings of Boxship 2001 – Future Evolution of the Containership, London

Vacca I, Bierlaire M, Salani M (2007) Optimization at container terminals: Status, trends and perspectives. In: Proceedings of the Swiss Transport Research Conference (STRC), Monte Veritá/Ascona, pp 1–21

Vis IFA, de Koster R (2003) Transshipment of containers at a container terminal: an overview. European Journal of Operational Research 147(1):1–16

Wang F, Lim A (2007) A stochastic beam search for the berth allocation problem. Decision Support Systems 42(4):2186–2196

Wiegmans BW, Rietveld P, Nijkamp P (2001) Container terminal services and quality. Serie Research Memoranda 0040, Free University Amsterdam

Wilson ID, Roach PA (2000) Container stowage planning: a methodology for generating computerised solutions. Journal of the Operational Research Society 51(11):1248–1255

Winter T (1999) Online and real-time dispatching problems. PhD thesis, Technical University Braunschweig

Zhang H, Kim KH (2009) Maximizing the number of dual-cycle operations of quay cranes in container terminals. Computers and Industrial Engineering 56(3):979–992

Zhou P, Kang H, Lin L (2006) A dynamic berth allocation model based on stochastic consideration. In: Proceedings of the 6th World Congress on Intelligent Control and Automation (WCICA 2006), IEEE Computer Society, Washington DC, vol 2, pp 7297–7301

Zhu Y, Lim A (2006) Crane scheduling with non-crossing constraint. Journal of the Operational Research Society 57(12):1464–1471

GPSR Compliance

The European Union's (EU) General Product Safety Regulation (GPSR) is a set of rules that requires consumer products to be safe and our obligations to ensure this.

If you have any concerns about our products, you can contact us on ProductSafety@springernature.com

In case Publisher is established outside the EU, the EU authorized representative is:

Springer Nature Customer Service Center GmbH
Europaplatz 3
69115 Heidelberg, Germany

Batch number: 10725802

Printed by Printforce, the Netherlands